THE CICINDELIDAE OF CANADA

The bright colours and fascinating ways of this small but important group of insects attract immediate attention. Cicindelidae, or tiger beetles, are frequently encountered, but they are difficult to capture, since they are alert and elusive, and still more difficult to identify. This intensive study of the distinguishing characteristics, geographical distribution and variation, and habits and habitats of tiger beetles in Canada—the culmination of the author's main interest for many years—will provide a much needed reference work. Studies of insect families are scarce, and professional and amateur entomologists alike will find this book a most useful aid in their investigations and a stimulus to further research.

J. B. WALLIS, one of Canada's most distinguished amateur entomologists, is an honorary member of the Entomological Society of Manitoba and was one of the founders of the Natural History Society of Manitoba, which awarded him its medal for outstanding work in entomology.

The Cicindelidae
of Canada

By

J. B. WALLIS

Winnipeg, Canada

UNIVERSITY OF TORONTO PRESS

University of Toronto Press

Diamond ◇ Anniversary 1961

*To my beloved wife Edna
and to my lifelong companions
the Criddles of Aweme, Manitoba
this book is dedicated*

PREFACE

ONE of the world's great biologists, J. B. S. Haldane, was once asked what one could conclude about the nature of the Creator from a study of His creation. The reply was "an inordinate fondness for beetles." This remark is based upon the fact that the beetles are the largest Order of living animals. Now, if the Creator was fond of beetles, He must have been especially fond of the Cicindelidae or tiger beetles, because they are formed so beautifully, coloured so brilliantly, and move so gracefully. These insects, some of which are illustrated in the Plates, are found throughout the warmer parts of the world, including the southern half of Canada.

This book summarizes what I have been able to learn about Canadian tiger beetles in a lifetime of which only the leisure hours have been devoted to the study of natural history. Many happy hours have been spent with these insects, both in the field and in my home, and I hope this volume will make it possible for others to share the pleasure that the study of tiger beetles has given to me. The keys and illustrations presented in the pages which follow make it possible to identify the Canadian Cicindelidae, and the maps show in detail where each species has been found. However, identification is only the beginning of study, not its end, and, in the text, I have indicated some of the problems related to the natural history of these beetles which require solution. We have a tremendous amount to learn about the species before we can say they are properly understood.

J.B.W.

ACKNOWLEDGMENTS

ALTHOUGH this work appears under the authorship of one man, it could not have been completed without the encouragement and co-operation of many others.

The following loaned or gave me specimens in their care: Dr. R. C. Froeschner, Montana State College; Mr. C. A. Frost, Framingham, Massachusetts; Dr. M. H. Hatch, University of Washington; Mr. Hugh B. Leech, California Academy of Sciences; Professor George Spencer, University of British Columbia; Drs. E. H. Strickland and G. E. Ball, University of Alberta; Drs. F. A. Urquhart and G. B. Wiggins, Royal Ontario Museum; and the Officers-in-Charge of the Dominion Entomology Laboratories at Saskatoon, Saskatchewan, and Lethbridge, Alberta.

Mr. W. J. Brown of the Entomology Research Institute, Ottawa, sent me paratypes of *Cicindela formosa fletcheri* Criddle. He helped me also with suggestions concerning the modern concept of species, and was kind enough to take time to prepare and send to me a list of the species of *Cicindela* from Ontario and Quebec contained in the Canadian National Collection. In addition, he sent me copies of species descriptions not otherwise available.

Dr. F. F. Hasbrouck of the Oregon State College at Corvallis compared specimens with paratypes of *Cicindela chamberlaini* Knaus. Dr. E. B. Britton, of the British Museum (Natural History) compared specimens with Kirby's type of *Cicindela albilabris*, besides obtaining evidence of Le Conte's identifications of *Cicindela longilabris* Say and *Cicindela montana* Le Conte.

Dr. Mont A. Cazier and Mrs. Patricia Vaurie of the American Museum of Natural History were particularly kind in advising me on the correct identification of some variable species.

Mr. A. R. Brooks of the Entomology Laboratory, Canada Department of Agriculture, at Saskatoon, not only took me to many interesting collecting localities in Saskatchewan but himself collected most successfully on my behalf.

To Professor A. G. Robinson of the University of Manitoba special thanks must go. He undertook the onerous task, not only of reading the manuscript critically, but of running through the key every form mentioned in this work.

Without the help of Dr. Ralph Bird, Head, Entomology Laboratory, Canada Department of Agriculture Research Station, Winnipeg, Manitoba, this book could not have been written. He suggested the project, gave encouragement through the difficult years of preparation, and arranged for editing and final publication. Not only did he allow me to study the collection in the Brandon Entomology Laboratory which contained some of the Criddle material, but he took all the photographs for the figures, acted as intermediary in obtaining some literature which is not easily accessible, and collected a very fine series of *Cicindela purpurea* Olivier, from near Brandon.

All credit for production of the positives for the fine coloured plates which grace this publication goes to Mr. G. H. Parker, Head, Biographic Unit, Canada Department of Agriculture, Ottawa.

Dr. A. J. Thorsteinson, Department of Entomology, University of Manitoba, kindly gave me permission to work in the Insect Museum of the University, to enlist the aid of the students in testing the key, and to receive the aid of the Department's stenographer in typing much of the manuscript.

Dr. G. E. Ball,* Department of Entomology, University of Alberta, spent much valuable time in reviewing and editing the manuscript. All references to genitalia are his. These include the suggested groups of species as well as some of the synonymy.

Miss McCowan of the staff of the Brandon Entomology Laboratory performed the tedious task of typing the preliminary drafts of this manuscript. She did a magnificent job, and I am glad to offer her my thanks and compliments. Miss Ollie Andrayko, Department of Entomology, University of Manitoba, prepared a revised copy of the manuscript. The final draft was prepared by Miss Joan Shore, Department of Entomology, University of Alberta. The maps were prepared by Mr. R. A. Brust, a graduate student, Department of Entomology, University of Manitoba.

To all the above I extend my sincere thanks.

J.B.W.

*Mr. Wallis, because of his advanced years and failing health, did not feel up to seeing his work through the press, and he entrusted me with this task. In fulfilling this responsibility, I have worked closely through correspondence with Miss Jean C. Jamieson, Associate Editor, University of Toronto Press. It is a pleasure to record my indebtedness to her for outstanding efforts in preparing the manuscript for publication. G.E.B.

CONTENTS

MAPS

THE CICINDELIDAE OF CANADA

I. Taxonomic Categories and Classification

The Cicindelidae belong to the sub-order Adephaga, and may be known by the following characteristics: eyes large, prominent; antennae filiform, comprising eleven articles, the outer seven of which are pubescent, inserted on the frons above and between the bases of the mandibles; metasternum with an antecoxal piece; legs cursorial; male genitalia with long, slender, glabrous, lateral lobes, joined by a narrow sclerotized yoke which passes over dorsal surface of median lobe; stylus of coxite of female retractile plate (or genitalia or ovipositor) toothed. The larva has chewing mouthparts, a flattened head, and the fifth abdominal tergum bears a pair of prominent hooks. The tiger beetles are very similar to the family Carabidae, and some authors include these two groups in a single family.

Horn (1916) lists the Cicindelidae of the world as including five subfamilies containing thirty-five genera and 1,299 species. Of the thirty-five genera, only four are recorded from North America, and of these, *Amblycheila* and *Omus* are peculiar to this continent. A third genus, *Megacephala*, is a large group in the American tropics; and the fourth genus, *Cicindela*, is world-wide. Two of these four genera— *Omus* and *Cicindela*—occur in Canada. The former is represented here by two species, the latter by twenty-six.

Although the majority of Canadian tiger beetles are readily identified there are some whose placing remains in doubt. For instance, the relationships among the populations of *Cicindela oregona* and *duode-cimguttata*, and even more among the forms of the *longilabris* group present an interesting challenge to Canadian coleopterists. Perhaps, then, some remarks on taxonomy are called for.

In the early days of classification of plants and animals, the term "species" meant only the members composing a local population of a "kind" of plant or animal. Such members had to conform in morphological characters very closely to a "type" specimen selected and so designated by the describer of the species, and the species itself was considered to be "fixed" or invariable. The theories of Darwin and Wallace, followed later by the rediscovery of the Mendelian rules of heredity, have gradually led to the discarding of the static concept of a species. The modern, commonly accepted definition of species whose reproduction is bisexual is as follows: a group of actually or potentially interbreeding natural populations which are reproductively isolated from other such groups. This definition means that if two species live in the same area, there will be no interbreeding between them, that is, each group will "breed true," and hybrids or specimens with intermediate characters will not be produced. Therefore, in determining the taxonomic status of recognizably different individuals collected in the same general area, one attempts to find out if intermediate types are also around: if so, then the variants are considered to be conspecific, because the intermediate type suggests that the extremes are capable of interbreeding; if not, the different forms are considered to represent distinct species.

Populations which occupy the same general area are said to be sympatric; populations which occur in adjacent areas are allopatric. The taxonomic status of similar, allopatric populations is often problematical, and a decision must be based on arbitrary criteria unless breeding experiments are conducted. If all of the individuals collected in one area are clearly distinct from all of the individuals in a separate area, then the two populations are considered to represent different species; if this distinction in appearance does not hold, then the geographically distinct populations are regarded as conspecific. If at least 75 per cent of the individuals from one area are separable from 75 per cent of the individuals from a different area, the two groups are said to represent different subspecies. Hence a subspecies is a geographically defined aggregate of local populations which differs taxonomically from other such subdivisions of the species. A species which is not divided into subspecies is monotypic. Contrasted with this is a polytypic species, which comprises two or more subspecies.

In the past, other categories to describe variation have been recognized. The commonest are "variety" and "aberration." They are usually applied to intra-population variants which are relatively infrequent or seasonal in occurrence. Such variants have no taxonomic status, and it is best not to apply formal names to them.

In some cases, individuals of two biologically distinct species are so similar morphologically that application of a morphological species definition leads to the conclusion that the two are conspecific. Species pairs such as these are said to be "sibling" (Mayr, 1942). In Canadian tiger beetles the subspecies of *Cicindela sexguttata* are a possible example: the two forms are not very different morphologically but they differ considerably in their ecological requirements, so perhaps they are sibling species.

For a more detailed discussion of the species problem from the taxonomic viewpoint, see Mayr *et al.* (1953).

II. The Biology of Tiger Beetles

LIFE HISTORY

CONSIDERABLE INVESTIGATION HAS TAKEN PLACE on the life history of *Cicindela*. Shelford and his associates (1908) were prominent in the study of the North American species. In Canada, Criddle (1910) did some excellent work on the species inhabiting the sand dunes of Manitoba. The observations of these workers may be summarized. Eggs are deposited singly in the soil, each in a burrow, which the female digs with her ovipositor. When the egg hatches, the newly emerged larva proceeds to enlarge the burrow, and lives in it. Before each moult, the larva closes the burrow, goes to the bottom, moults, and reappears in about one week. In the fall the burrow is closed, and the larva hibernates, usually below the frost line. There are only three larval instars in the life cycle, passed in the course of two or three years. Pupation takes place in an especially prepared chamber, usually at one side of the main burrow.

The Adults

Little is known of the biology of the genus *Omus*. The species are nocturnal and may be found in daytime under cover in fields and open forests.

Most of the species of the genus *Cicindela* fly well and all can run, some exceptionally fast. The adults are very sensitive to the stimuli of heat and light. Although they are creatures of the sun, excessive heat will drive them to the shelter of vegetation or of a burrow.

Chapman *et al.* (1926) showed that on the sand dunes of Minnesota a temperature of about 125° F. is lethal for *Cicindela formosa*, and that *lepida* can stand only about 118° F. These temperatures often occur in the micro-climate of sand dunes and give a ready explanation for the scarcity of tiger beetles during the heat of the day. No doubt a much lower temperature is lethal for some species, for instance *sexguttata*, a dweller on the cool floor of deciduous forest.

Response to changes in light is very noticeable, for even the shadow of a cloud will check activity. Although *Cicindela* are pronouncedly diurnal, many species can be attracted on suitable nights by a light placed on the ground over which they run during the day. The Vauries (1950) found this to be a profitable way of securing several species considered to be rare.

All of our adult *Cicindela* are known to, or believed to, dig short burrows in which to pass the night, or to shelter from excessive heat or moisture. These short burrows are easily and rapidly constructed, but the deep burrows for overwintering are evidently a major engineering work. The latter burrows vary in depth from 6 inches (*repanda*) to 48 inches (*formosa*) depending on the species and on the type of soil: burrows dug in clay are shallower than those dug in sand. The winter burrows of the larva and adult of the same species are approximately of the same depth.

In preparing the winter burrow, the adult first loosens the soil with its mandibles, and then uses one of its front legs to scrape the soil into a heap beneath its body. It uses one of its middle legs to pass the soil further back, and then one of its hind legs to kick the soil clear. For some inches down what is to be the burrow, the earth is thrown out, but from then on the soil is packed behind the beetle as it descends; the burrow is kept large enough for its inmate to be able to turn around. When all is prepared satisfactorily, the beetle goes to the extreme end, turns around so that its head is upward, and hibernates in this position. It has left room at the bottom of the burrow to contain the earth which it will push down from above when it climbs out the following year.

Adult *Cicindela* may be found on almost any bare spot of soil where the surface has not been disturbed or trampled, and each species has a marked preference for a certain type of soil. Roughly, the Canadian species may be divided into three groups, according to their preferences: (1) those preferring soil in which sand is predominant; (2) those preferring gravelly or sandy loam to clay; (3) those preferring alkaline or saline soil. To some species, slope is of great importance, and some prefer banks of rivers or lakes. Some species seem to require the presence of sparse vegetation; others prefer shady places.

These insects are carnivorous, and they actively hunt, capture, and eat any living creature not too large for them. They have been reported eating moss or algae, but this may have been only a way of securing a water supply. Ants make up a considerable proportion of their diet (and also of the diet of the larvae), but they are not taken with impunity; it is not uncommon to find an ant's head tightly fastened to a tiger beetle's antenna or its front leg.

An interesting account of tiger beetles as predators is given by Frick (1957). Adults of *haemorrhagica* Le Conte and *pusilla imperfecta* Le Conte were found to be feeding on larvae and adults of the alkali bee, *Nomia melanderi* Cockerell. This bee is important in the pollination of alfalfa, and so a chemical control programme was instigated to reduce the numbers of predatory tiger beetles. Later it was discovered that the predations of the tiger beetles had little effect on the numbers of alkali bees. One is inclined to ask if it would not have been better to have assessed first the economic importance of the beetles.

If tiger beetles are predators on some creatures, they are also the prey of others. A bombyliid fly, *Spogostylum anale* Say, is an external parasite on the larva of *C. scutellaris lecontei*, and I have on two occasions seen a large robber fly capture an adult tiger beetle on the wing, and on another occasion found a smaller asilid eating an adult *limbata nympha*. Presumably the robber flies can catch the beetles only when the latter are in flight with the vulnerable part of their bodies exposed between the wing covers, because the beak of the robber fly is inserted just behind the scutellum.

Many species of *Cicindela* are so coloured and marked that they blend remarkably with their surroundings. A notable example is *lepida*, a dweller on bare sand, whose sand-coloured elytra with their vague dark markings make it very difficult to see when motionless and near the observer's feet. Indeed it is often easier to see the shadow than the beetle itself.

The Larvae of Cicindela

The larvae are beautifully adapted for life in burrows. The jaws are powerful and bent upward. The head and pronotum are bent downward at an angle of about 45°, and are strongly sclerotized above; together they form a somewhat rounded plate. The underside of the head is sclerotized and swollen. The S-shaped body is very flexible, and is well provided with short, strong spines. On the fifth abdominal tergum is a hump, bearing some forward-curving hooks. The tarsal claws are long and sharp.

In enlarging its burrow the larva uses its jaws to loosen and dig the soil and then shovels it aside with the head and pronotum. As the burrow increases in depth, the larva turns around with its load of earth, works its way up to the top, and then, with the help of the rounded plate formed by the head and pronotum, throws the load of earth away, perhaps to a distance of as much as 3 inches. It uses the underside of its head to pack the walls of the burrow and the area around the entrance, which must always be kept clear and firm. When the larva wants to plug the entrance to its burrow, it uses the rounded plate formed by its head and pronotum.

Normally, the larva lies in its burrow with its jaws extended to seize small insects that venture within reach. It will eat almost any insects that are small enough for it to handle; it shows a preference for ants but a distaste for Hemiptera. Should it seize something that threatens to pull it out of its burrow, it immediately brings its defensive mechanism into play. The tarsal claws take firm hold of the soil, and the body flexes so as to force its spines and the curved hooks of the hump into the walls of the burrow. If, in spite of all this pressure, the enemy is evidently too strong to be captured, the larva releases the hold of its jaws, retracts its spines and curved hooks from the walls, and literally drops to the bottom of the burrow.

The depths to which the winter burrows descend are amazing. Criddle found that the third-stage larvae of *C. lepida*, a small species, descended to as much as 6 feet. The larger larvae of *C. formosa manitoba* went down an average distance of 66 inches, the greatest distance measured being 78 inches.

III. The Genera of Canadian Tiger Beetles

THE ADULTS OF THE TWO CICINDELID GENERA which occur in Canada may be distinguished from one another as follows:

Eyes small; posterior coxae plainly separated   *Omus* Eschscholtz
Eyes large; posterior coxae contiguous   *Cicindela* Linnaeus

The Genus Omus Eschscholtz, 1829

Once a specimen of *Omus* has been seen there is no likelihood of any confusion between the genera. This genus is restricted to the Pacific Coast of North America; in Canada, it is found only on Vancouver Island and the adjacent mainland of British Columbia. The two species are readily distinguished from one another as follows:

Elytra conspicuously foveate   *O. dejeani* Reiche
Elytra not foveate   *O. californicus* Eschscholtz

Omus (Megomus) dejeani Reiche, 1838 (No. 1, Plate 1)

Colour black or blackish-brown, length 15–21 mm.

This species is said to be sometimes found in numbers by breaking up rotton trunks of trees, and Schaupp (1883) states that it may be taken by baiting it with finely chopped meat placed under boards, where the insects may be found on the day following. The beetles must be handled with care for they are quick in their movements and bite readily and viciously.

This species is found in western British Columbia, western Washington, western Montana, and California. In British Columbia, it has been collected in the following localities: Eberts, New Westminster, and Pender Harbour on the mainland; and Royal Oak, Saanich District, on Vancouver Island.

Omus (Omus) californicus Eschscholtz, 1829 (No. 2, Plate 1)

Colour black, length 13–18 mm.

This species occurs in dry situations under stones, and is rare. The species as a whole ranges from southwestern British Columbia nearly to Los Angeles, California. In Canada, where it is represented by the subspecies *c. audouini* Reiche, 1838, it has been recorded from Vancouver Island and the adjacent mainland.

The Genus Cicindela Linnaeus, 1758

Of the twenty-six species of *Cicindela* which inhabit Canada, ten are monotypic and sixteen are polytypic. Five of the polytypic species are represented in this country by one subspecies each, and the remaining eleven species by thirty-one subspecies. Thus the total number of named forms for consideration is forty-six.

Some attempts have been made to divide this genus into genera or at least subgenera. For example, Rivalier (1954) reclassified the species on the basis of structure of the male genitalia. If his scheme were followed, the Canadian species would be arrayed in four genera and three subgenera. Papp (1952), who also studied the male genitalia, divided the North American species into four groups, which did not correspond to Rivalier's segregates. The male genitalia of the North American species have been studied by G. E. Ball. I have seen his classification, and it is at variance with those of the authors cited above. I accept his opinion that all of the North American species should be included in a single genus, and I also accept his arrangement of the species, which is used here. However, I do not think that the inter-relationships of the species are settled as yet, so I do not recognize any of the proposed groupings, including those of earlier authors.

Proper identification of a tiger beetle depends upon the ability of the worker to interpret the couplets in the key which follows. To ensure that this key will be used correctly, some of the characters used in it are here explained briefly.

The principal characteristics used to identify species of *Cicindela* are: presence or absence of hair on the face and dorsal surface of the head; presence or absence of teeth on the elytral apices; and markings of the elytra.

The entire frons and vertex of the head may be hairy (Fig. 1); or the hairs may be restricted to a small patch near the front inner edge of each eye (Fig. 2); or hair may be lacking from the frons and vertex, except for the usual long, widely spaced bristles along the orbits of the eyes (Fig. 3).

The apices of the elytra must be examined carefully to determine if they are serrulate or not. The edges of the elytra must be clean, and should be viewed in silhouette through a microscope or hand lens, preferably against a white background. A piece of white paper can be inserted between the tips of the elytra and abdomen, and this will facilitate examination. The serrations begin about where the edge of each elytron curves toward the apex of the suture. They continue almost to the suture toward which they are most prominent (Fig. 4). In some species, this character varies geographically, so that some populations have serrate elytra and others have smooth elytra. Such a species is placed in two divisions of the key.

Although it is true that the markings of many species of *Cicindela* vary greatly intra-specifically, yet the type and degree of the markings are much used in identification, and therefore a clear conception of them is necessary. The pale portions of the elytra are referred to as the maculation or markings, although actually it is only the dark areas that are pigmented. These pale markings are composed of spots and lines which have received names.

Beginning at the base of the elytron, just inside the shoulder, a pale line runs outward to the point of the shoulder, and thence down along the margin for a short distance before curving or turning on to the disk of the elytron. This marking is known as the *humeral lunule* and three of its shapes are shown in No. 46, Plate 2, and Nos. 50 and 56, Plate 3.

Extending inward from the margin about the middle of each elytron is a transverse line which bends backward into a longish extension. This is the *middle band* and may be seen on each of the coloured illustrations cited above.

At the apex of the elytron and beginning at the suture is a line which extends outward along the apical margin and upward for a short distance before turning on to the elytral disk. This is the *apical lunule* and may be seen in the same coloured illustrations.

Along the margins, beginning at the outer end of the middle band, is a line extending towards the shoulder and down towards the apex. This line may be quite short or may join both the apical and the humeral lunules. This is the *marginal line* or *band*, and it, too, is plainly shown in the same coloured illustrations.

When all these markings are present and entire they are said to be *complete*. However, they are subject to much disintegration, some even disappearing, and all or some may be represented by dots. No. 66, Plate 3, shows a specimen from which some of the markings have disappeared, and No. 91, Plate 4, one in which only a few dots remain. Such markings are said to be *incomplete*.

Key to the Canadian Species of Cicindela Linnaeus

1. Head without hairs on frons or between eyes, other than a few widely spaced long bristles usually found along orbit of each eye (Fig. 3)   3

    Head with a cluster of hairs near inner edge of each eye, or all over frons   2

2. (1) Head with a cluster of hairs near anterior inner edge of each eye, no hair on remainder of space between and behind eyes and frons (Fig. 2)   9

    Head with frons and excavation between and behind eyes more or less hairy (Fig. 1)   10

3. (1) Elytral apices not serrulate   4

    Elytral apices serrulate (Fig. 4)   6

4. (3) Labrum elongate (Nos. 60–69, Plate 3)   5

    Labrum not elongate (Nos. 9–10, Plate 1)   *pusilla* Say, p. 61.

5. (4) Surface of elytra shining between punctures and foveae
    montana Le Conte, p. 49.

    Surface of elytra more or less opaque, not or only slightly shining, with many minute granules   *longilabris* Say, p. 46.

6. (3) Colour of elytra brown-bronze to greenish; markings complete and heavy (No. 76, Plate 3)   *willistoni* Le Conte, p. 60.

    Colour varied; markings never complete, often wanting or reduced to spots   7

7. (6) Form elongate, narrow for genus; colour blackish-brown or slightly greenish, with a bronzy sheen; a row of bluish or greenish foveae on each elytron (No. 96, Plate 4)
    punctulata Olivier, p. 61.

    Form broader, as usual for genus; colour of elytra blue or green; no dorsal foveae on elytra (Nos. 91–95, Plate 4)   8

8. (7) Colour shining and quite brilliant; maculation varying from completely absent to eight to ten dots; the middle band very rarely complete (Nos. 91–94, Plate 4)
    sexguttata Fabricius, p. 30.

    Colour duller; humeral lunule of two dots or very occasionally complete; middle band complete; apical lunule of two dots (No. 95, Plate 4)   *patruela* Dejean, p. 33.

9. (2) Elytra not punctate; elytral apices not serrulate; labrum black or dark (Nos. 84–90, Plate 4)   *scutellaris* Say, p. 33.

    Elytra punctate; elytral apices serrulate or not; labrum normal   *oregona* Le Conte, p. 22.

10. (2) Hair on head and body lying flat   11
    Hair on head and body more or less erect   14
11. (10) Ground colour of elytra yellow   12
    Ground colour of elytra brownish   13
12. (11) Dark markings of elytra not sharply defined, bronzy or
    bronzy-green (No. 106, Plate 4)   *lepida* Dejean, p. 67.
    Dark markings of elytra one or two rounded or longitudinal
    spots and a basal triangular scutellar area with its apex usually
    extended along suture as a line to elytral apices *limbata* Say, p. 27.
13. (11) Each elytron with a whitish spot between scutellum and
    humerus (No. 105, Plate 4)   *cuprascens* Le Conte, p. 65.
    Each elytron without a spot between scutellum and humerus
                                    nevadica Le Conte, p. 67.
14. (10) Elytral apices not serrulate, or sometimes very feebly so
    (cf. also *scutellaris* Say, couplet 9)   15
    Elytral apices distinctly serrulate   22
15. (14) Humeral lunule absent, *or* a feeble projection, *or* one or
    two dots   16
    Humeral lunule long and oblique, *or* a short spur, *or*
    obliterated by marginal band; markings heavier and more
    nearly complete   20
16. (15) Elytra immaculate, *or* with a few spots, *or* with a
    marginal band, the middle band represented by never more
    than a dot or a short projection from marginal band; elytra
    feebly punctured at most, and scarcely granulate (Nos. 84–
    90, Plate 4)   *scutellaris* Say, p. 33.
    Elytra with incomplete markings, but always with at least
    a partial middle band the inner arm of which is oblique;
    elytra not strongly punctured, but evidently granulate (Nos.
    27–36, Plate 2)   17
17. (16) Middle band of elytron arising from complete marginal
    band (No. 37, Plate 2)   *cimarrona* Le Conte, p. 39.
    Middle band of elytron separated from marginal band   18
18. (17) Middle band of elytron rather widely separated from
    margin, the short basal part (sometimes absent) somewhat
    oblique, the downward arm short and oblique; post-humeral
    dot usually absent (Nos. 27–33, Plate 2) *purpurea* Olivier, p. 43.
    Middle band of elytron narrowly separated from margin,
    the basal portion longer and more transverse, the downward
    arm longer; post-humeral dot usually present   19

19. (18) Downward arm of middle band of elytron not unusually
    long (Nos. 34–38, Plate 2)   *limbalis* Klug, p. 42.
    Downward arm of middle band of elytron unusually long,
    all markings heavy, but not complete; colour of elytra
    greenish, sometimes with a feebly cupreous sheen (No. 38,
    Plate 2)   *decemnotata* Say, p. 40.
20. (15) Marginal line rarely complete, sometimes absent, un-
    usually short and narrowed at each end; distal end of humeral
    lunule long and oblique (Nos. 51–59, Plate 3)
                                    tranquebarica Herbst, p. 57.
    Marginal line complete and unusually heavy, sometimes
    narrowing where it joins humeral lunule; humeral lunule
    obliterated by expansion of humeral line (No. 12, Plate 1),
    or distal end of lunule showing as a short unusually sub-
    conical spur (Nos. 4–8, Plate 1), *or* the distal end an evident,
    rather heavy, oblique line (No. 17, Plate 2)   21
21. (20) Pale markings moderately to very wide, the humeral
    lunule either completely obliterated by extension of marginal
    band, *or* represented by a short spur, *or* normally developed
    (Nos. 3–16, Plate 1); labrum not much produced
                                         formosa Say, p. 36.
    All pale markings well developed but marginal line not
    expanded, the humeral lunule long, its oblique spur often
    nearly touching middle band (Nos. 17–21, Plate 2)
                                      lengi W. Horn, p. 54.
22. (14) Pale areas of elytra reduced, the humeral lunule absent
    or represented by never more than two dots   23
    Pale markings of elytra more complete and heavier, the
    humeral lunule always represented by a complete marginal
    line when not obliterated by extension of pale margin   28
23. (22) Colour brownish-black, the marking of elytra some-
    times broken into dots; middle band, when present, entering
    from margin nearly at right angles, its inner arm descending,
    also nearly at right angles; sometimes this band expands
    briefly along margin (Nos. 40–46, Plate 2; Nos. 47–48, Plate
    3)   24
    Colour cupreous, cupreous-purple, green- or blue-purple,
    green, blue, or black, the colours always somewhat metallic;
    inner arm of middle band oblique   17
24. (23) Labrum with three, more or less acute, teeth (No. 39,
    Plate 2)   *ancocisconensis* Harris, p. 57.

Labrum with one acute tooth, the two lateral teeth represented by rounded projections   25
25. (24) Elytral markings reduced or partially obliterated   26
    Elytral markings complete   27
26. (25) Elytral markings much broken, humeral and apical
    lunules divided into spots or absent; middle band reduced
    and marginal line at most short; pronotum short and scarcely
    convex (Nos. 40–44, Plate 2)

        typical *duodecimguttata* Dejean, p. 20.

    Elytral markings varying from six dots to usual *repanda*
    pattern; colour light bronze; form and pronotum as in *re-
    panda* (Nos. 47–48, Plate 3)   *repanda novascotiae* Vaurie, p. 18.
27. (25) Colour of elytra blackish-brown; form broader and
    flatter than in *repanda*; white side-margins of elytra widely
    separated from humeral lunule (No. 41, Plate 2)

        atypical *duodecimguttata* Dejean, p. 21.

    Colour of elytra brown-bronze, sometimes with greenish or
    coppery reflections; white side-margins of elytra nearly or
    quite reaching humeral lunule (Nos. 45 and 46, Plate 2)

        r. repanda Dejean, p. 18.

28. (22) Humeral lunule obliterated by expansion of marginal
    line (No. 23, Plate 2), *or* distal end of humeral lunule showing
    as a short, usually subconical spur (No. 4, Plate 1); *or* distal
    end an evident, rather heavy, oblique line (No. 19, Plate 2)   29

    Distal end of humeral lunule with inner edge curved or
    angled, forming a more or less distinct inverted C (Nos. 49
    and 50, Plate 3)   32
29. (28) Labrum with three teeth   30
    Labrum with one tooth   31
30. (29) Surface of elytra appearing greasy, length usually less
    than 13.0 mm. (Nos. 77–82, Plate 4)   *fulgida* Say, p. 51.
    Surface of elytra without a greasy appearance   20
31. (29) Surface of elytra greasy in appearance; markings com-
    plete, colour mainly bright reddish (No. 83, Plate 4)

        parowana Wickham, p. 53.

    Surface of elytra not greasy in appearance; elytra mainly
    white, if markings nearly complete then colour is dark brown
    or black, not bright red (Nos. 22–26, Plate 2) *limbata* Say, p. 27.
32. (28) Distal end of humeral lunule entering elytral disc in a
    curve (No. 4, Plate 1)   24

    Distal end of humeral lunule entering elytral disc at an

angle, obtuse or nearly right, and with its extreme tip usually turned forward (Nos. 49–50, Plate 3)   33

33. (32) A white dot at base of each elytron between shoulder and scutellum (No. 105, Plate 4)   *cuprascens* Le Conte, p. 65.

No white dot at base of elytron (No. 104, Plate 4)   34

34. (33) Pubescence decumbent, much like *cuprascens* in general appearance but smaller   *nevadica* Le Conte, p. 67.

Pubescence erect or decumbent (Nos. 49–50, Plate 3)

hirticollis Say, p. 25.

IV. The Canadian Species of *Cicindela* Linnaeus

Cicindela repanda Dejean, 1825
(Nos. 45–46, Plate 2, Nos. 47–48, Plate 3, and Map 1)

THIS SPECIES is readily confused with some specimens of *duodecim-guttata*. In areas where the two species are sympatric, the marginal band in *repanda* is not, or else very narrowly, interrupted, whereas in *duodecimguttata* this band is widely interrupted. However, if a positive identification is required, then the male genitalia must be dissected and studied (see Lindroth, 1955, 16–17).

GEOGRAPHICAL VARIATION AND SUBSPECIES

This species is remarkably constant in colour and maculation throughout most of Canada. In specimens from southeastern British Columbia the marginal band of the elytron is more widely interrupted. In Nova Scotia and some of the adjacent islands, but not in Newfoundland, populations occur which have reduced or partially obliterated elytral markings. The wide-ranging form is the subspecies *repanda repanda* (Nos. 45–46, Plate 2); the eastern coastal form is *repanda novascotiae* Vaurie (Nos. 47–48, Plate 3).

SYNONYMY

W. Horn (1930) lists *bucolica* Casey as a synonym of *repanda*. This appears to be incorrect as it is possible to trace an unbroken series from typical *duodecimguttata* to the most completely marked *bucolica*. Furthermore, the shape of the pronotum is different in *repanda* and *bucolica*, the latter resembling *duodecimguttata*; and the habitats of *bucolica* are those of *duodecimguttata*, not of *repanda*.

RELATIONSHIPS

The external characteristics, and more especially the form of the male genitalia, indicate that this species and *duodecimguttata, oregona, hirticollis,* and *limbata* form a natural group. Probably *duodecimguttata* Say is the closest relative of *repanda*.

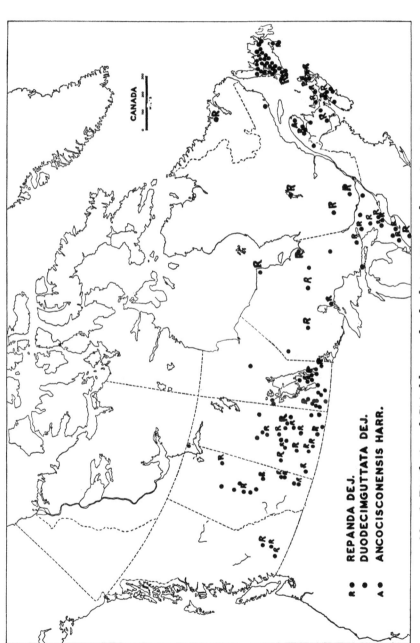

MAP 1. Distribution of *Cicindela repanda, duodecimguttata,* and *ancocisconensis*

REPANDA DEJ.

DUODECIMGUTTATA DEJ.

ANCOCISCONENSIS HARR.

CANADA

HABITAT

Cicindela repanda is probably the commonest tiger beetle in Canada. It is found on reasonably heavy, moist soil along river banks, especially on mud flats and sandy bars up to a foot or so from the water's edge. Although *duodecimguttata* occurs in the same general areas, it appears to be dominant above the zone occupied by *repanda*, and the two species usually do not mix.

LOCALITIES

Map 1 indicates the positions of the localities which are listed below.

repanda repanda Dejean: NEWFOUNDLAND, only at South Branch on Grand Codroy River; NEW BRUNSWICK, Bathurst, Boiestown; QUEBEC, Cascapedia, Godbout, Gracefield; ONTARIO, Algonquin Park, Atta-wapiskat, Black Sturgeon Lake, Cape Henrietta Maria, Don Valley, Glen Williams, Hopedale, Humber Bank, Hyde Park Corner, Leskard, London, Lake Nipissing, Macdiarmid, Musselman's Lake, Moose Factory, Nipigon, Newcastle, Normandale, Ottawa, Palm Lake, Point Pelee, St. Thomas, Sioux Lookout, Toronto, Ventnor, Willowdale, Woodstock; MANITOBA, Aweme, Elk Island, The Pas, Rennie, Shoal Lake, Victoria Beach, Waugh; SASKATCHEWAN, Beaver Creek, Christopher Lake, Elbow, Estevan, Fish Lake, Good Spirit Lake, Holbein, Lake Chaplin, Lumsden, North Battleford, Pike Lake, Rudy, Saskatoon, Torch River, Tunstall, Yonker; ALBERTA, Edmonton, Empress, Fawcett, Fort McMurray, Gull Lake, Lesser Slave Lake, Medicine Hat, Wainwright; BRITISH COLUMBIA, generally distributed especially on river beaches east of the Cascades

repanda novascotiae Vaurie: NOVA SCOTIA (including Magdalen Islands), Baddeck (Cape Breton), Great Village, Truro; PRINCE EDWARD ISLAND, Brackley Beach

Cicindela duodecimguttata Dejean, 1825
(Nos. 40–44, Plate 2, and Map 1)

This species is readily distinguished from all others except *repanda* by its rather subdued brownish colour and by the form of the markings of the elytra.

GEOGRAPHICAL VARIATION

In eastern Canada, the markings of the elytra of this species are more or less broken (No. 40, Plate 2). Farther west, especially in the southern portions of the prairie provinces, populations occur which

PLATES

PLATE 1

1. *Omus (Megomus) dejeani* Reiche, Victoria, B.C.
2. *Omus (Omus) californicus audouini* Reiche, Victoria, B.C.
3. *Cicindela formosa formosa* Say, Great Sand Dunes, Colo.
4. *Cicindela f. manitoba* Leng, Aweme, Man.
5. *Cicindela f. manitoba* Leng, Sand Hills s.w. Oak Lake, Man.
6. *Cicindela f. manitoba* Leng (cf. No. 11), Aweme, Man.
7. *Cicindela f. manitoba* Leng (cf. No. 10), Aweme, Man.
8. *Cicindela f. manitoba* Leng, Aweme, Man.
9. *Cicindela formosa generosa* Dej., Pelee Island, Ont.
10. *Cicindela f. generosa* Dej. (cf. No. 7), Chicago, Ill.
11. *Cicindela f. gibsoni* Brown (cf. No. 6), Pike Lake, Sask.
12. *Cicindela f. gibsoni* Brown, Elbow, Sask.
13. *Cicindela f. gibsoni* Brown, Elbow, Sask.
14. *Cicindela f. gibsoni* Brown, Pike Lake, Sask.
15. *Cicindela f. manitoba* Leng (underside), Lauder, Man.
16. *Cicindela f. gibsoni* Brown (underside), Beaver Creek, Sask.

PLATE 2

17. *Cicindela lengi lengi* W. Horn, Colorado Springs, Colo.
18. *Cicindela l. lengi* W. Horn (underside), Colorado Springs, Colo.
19. *Cicindela l. versuta* Csy., King's Park, Man.
20. *Cicindela l. versuta* Csy. (underside), Prince Albert, Sask.
21. *Cicindela l. versuta* Csy., Medicine Hat, Alta.
22. *Cicindela limbata limbata* Say, Wray, Colo.
23. *Cicindela l. nympha* Csy. (no subhumeral spot), Carberry, Man.
24. *Cicindela l. nympha* Csy. (subhumeral spot present), Elbow, Sask.
25. *Cicindela l. hyperborea* Le C., Northwest Territories
26. *Cicindela l. hyperborea* Le C., Northwest Territories
27. *Cicindela purpurea purpurea* Oliv., Exeter, N. H.
28. *Cicindela p. purpurea* x *p. auduboni*, Brandon, Man.
29. *Cicindela p. purpurea* x *p. auduboni*, Brandon, Man.
30. *Cicindela p. purpurea* x *p. auduboni*, Brandon, Man.
31. *Cicindela p. purpurea* x *p. auduboni*, Brandon, Man.
32. *Cicindela purpurea purpurea* x *p. auduboni* (*nigerrima* Leng), Brandon, Man.
33. *Cicindela p. pugetana* Csy., Vernon, B.C.
34. *Cicindela limbalis* Klug (typical), Algoma, Ont.
35. *Cicindela limbalis* Klug (*awemeana* Csy.), No. 1 Highway, Man.
36. *Cicindela limbalis* Klug (*spreta* Le C.), Nelson River, Man.
37. *Cicindela cimarrona* Le C., Jemez Mountains, N.M.
38. *Cicindela decemnotata* Say, Medicine Hat, Alta.
39. *Cicindela ancocisconensis* Harr., New Hampshire, U.S.A.
40. *Cicindela duodecimguttata* Dej. (typical), Manitoba
41. *Cicindela duodecimguttata* Dej. (*bucolica* Csy.), Lauder, Man.
42. *Cicindela duodecimguttata* Dej. (*hudsonica* Csy.), Ogoki, Ont.
43. *Cicindela duodecimguttata* Dej. (*hudsonica* Csy.), Ogoki, Ont.
44. *Cicindela duodecimguttata* Dej. (*hudsonica* Csy.), Ogoki, Ont.
45. *Cicindela repanda repanda* Dej., male, Elbow, Sask.
46. *Cicindela r. repanda* Dej., female, Treesbank, Man.

PLATE 3

47. *Cicindela repanda novascotiae* Vaurie, Centreville, Kings County, N.S.
48. *Cicindela r. novascotiae* Vaurie, Aldershot, Kings County, N.S.
49. *Cicindela hirticollis* Say, Berens Island, Man.
50. *Cicindela hirticollis* Say, Aweme, Man.
51. *Cicindela tranquebarica* Hbst. (typical), Florence, S.C.

52. *Cicindela tranquebarica* Hbst. (*minor* Leng), Mena, Ark.
53. *Cicindela tranquebarica* Hbst. (*borealis* Harr., det. Cazier), Salt Plain, N.W.T.
54. *Cicindela tranquebarica* Hbst. (*borealis* Harr., det. Cazier), Salt Plain, N.W.T.
55. *Cicindela tranquebarica* Hbst. (*Kirbyi* Le C., det. Cazier), Salt Plain, N.W.T.
56. *Cicindela tranquebarica* Hbst. (*kirbyi* Le C., det. Cazier), Westbourne, Man.
57. *Cicindela tranquebarica* Hbst. (*kirbyi* Le C., det. Cazier), Lake Como, Wyo.
58. *Cicindela tranquebarica* Hbst. (*kirbyi* Le C., det. Cazier), Ash Creek, Wash.
59. *Cicindela tranquebarica* Hbst. (*vibex* Horn), Kern County, Calif.
60. *Cicindela montana montana* Le C., Bozeman, Mont.
61. *Cicindela m. montana* Le C., Hill County, Mont.
62. *Cicindela m. spissitarsis* Csy., Calgary, Alta.
63. *Cicindela montana chamberlaini* Knaus, Lorna, B.C.
64. *Cicindela m. oslari* Leng, Wynndel, B.C.
65. *Cicindela m. laurenti* Schaupp, near Ward, Colo.
66. *Cicindela longilabris longilabris* Say, Sioux Narrows, Ont.
67. *Cicindela l. longilabris* Say, The Pas, Man.
68. *Cicindela l. novaterrae* Leng, Gander, Nfld.
69. *Cicindela l. novaterrae* Leng, Harmon Field, Nfld.
70. *Cicindela oregona oregona* Le C., Creston, B.C.
71. *Cicindela o. oregona* Le C. (underside), Creston, B.C.
72. *Cicindela o. oregona* Le C. (*scapularis* Csy.), Hope Trail, B.C.
73. *Cicindela o. guttifera* Le C., Echo, Utah
74. *Cicindela o. guttifera* Le C. (underside), Colorado Springs, Colo.
75. *Cicindela o. depressula* Csy., Mount Rainier, Wash.
76. *Cicindela willistoni echo* Csy., Owens Lake, Calif.

PLATE 4

77. *Cicindela fulgida fulgida* Say, Howard, Colo.
78. *Cicindela f. fulgida* Say (unusually large), southwestern Alberta
79. *Cicindela f. westbournei* Calder, Westbourne, Man.
80. *Cicindela f. westbournei* Calder, Roche Percee, Sask.
81. *Cicindela f. westbournei* Calder, Blucher, Sask.
82. *Cicindela f. westbournei* Calder, between Clavet and Elstow, Sask.
83. *Cicindela parowana wallisi* Calder, Penticton, B.C.
84. *Cicindela scutellaris scutellaris* Say, Colorado Springs, Colo.
85. *Cicindela s. scutellaris* Say, Borget, Tex.
86. *Cicindela s. scutellaris* Say, Cleveland County, Okla.
87. *Cicindela s. lecontei* Hald. (typical), Toronto, Ont.
88. *Cicindela s. lecontei* Hald. (typical), Aweme, Man.
89. *Cicindela s. lecontei* Hald. (*criddlei* Csy.), Treesbank, Man.
90. *Cicindela s. lecontei* Hald. (*criddlei* Csy.), Aweme, Man.
91. *Cicindela sexguttata sexguttata* Fab., Montreal, Que.
92. *Cicindela sexguttata sexguttata* Fab., Chicago, Ill.
93. *Cicindela s. sexguttata* Fab. (*harrisi* Leng), Clark County, Ind.
94. *Cicindela s. denikei* Brown, Betula Lake, Man.
95. *Cicindela patruela* Dej., Barbour, Ky.
96. *Cicindela punctulata punctulata* Oliv., Aweme, Man.
97. *Cicindela pusilla pusilla* Say, Westbourne, Man.
98. *Cicindela p. pusilla* Say, Stony Mountain, Man.
99. *Cicindela p. pusilla* Say, Hilton, Man.
100. *Cicindela p. pusilla* Say, Hilton, Man.
101. *Cicindela p. imperfecta* Le C., Cariboo, B.C.
102. *Cicindela p. cinctipennis* Le C., Medicine Hat, Alta.
103. *Cicindela nevadica knausi* Leng, Hilton, Man.
104. *Cicindela n. knausi* Leng, Hilton, Man.
105. *Cicindela cuprascens* Le C., Optima, Okla.
106. *Cicindela lepida* Dej., Aweme, Man.

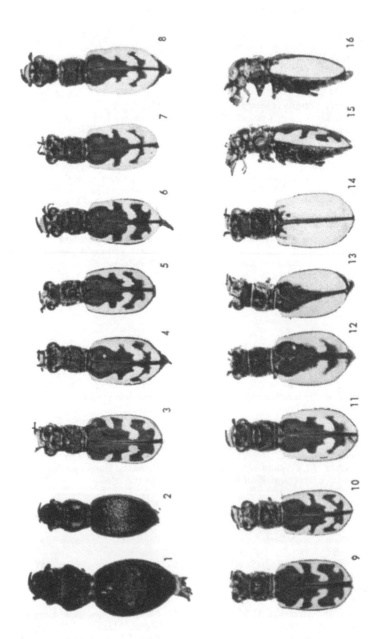

PLATE 1

PLATE 2

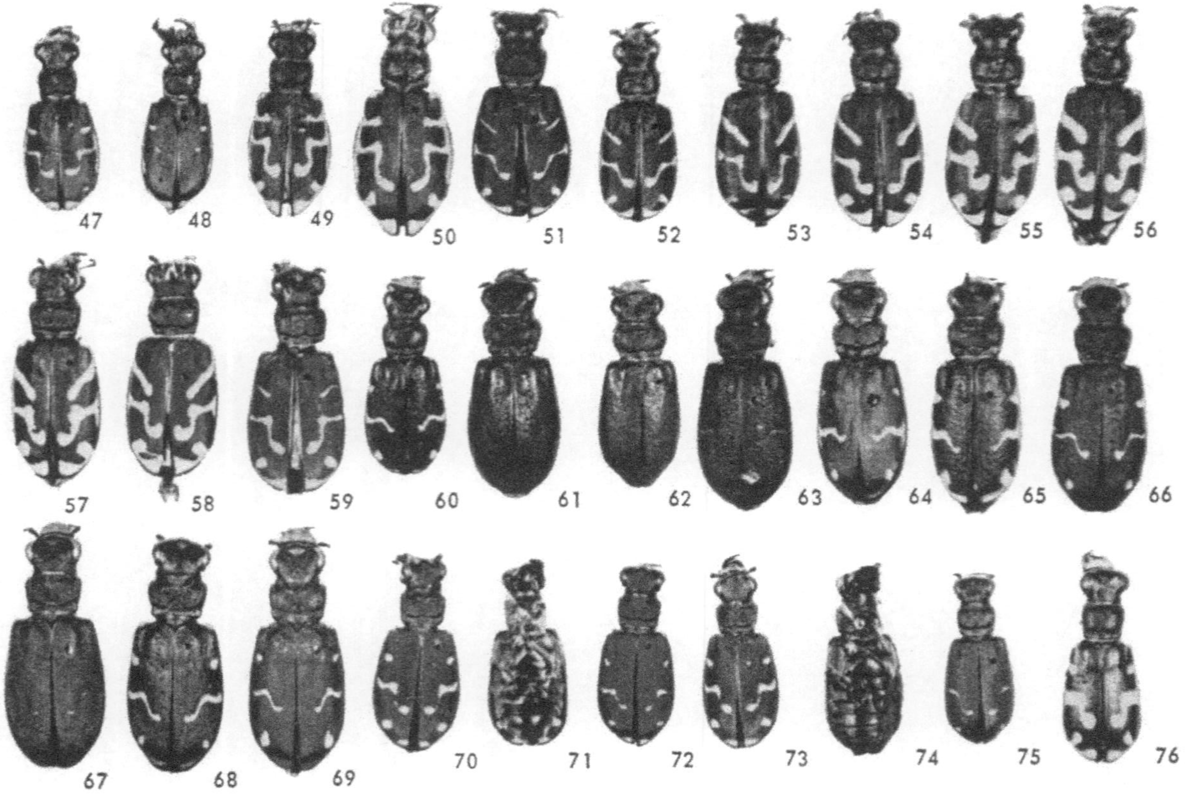
47
48
49
50
51
52
53
54
55
56
57
58
59
60
61
62
63
64
65
66
67
68
69
70
71
72
73
74
75
76

PLATE 3

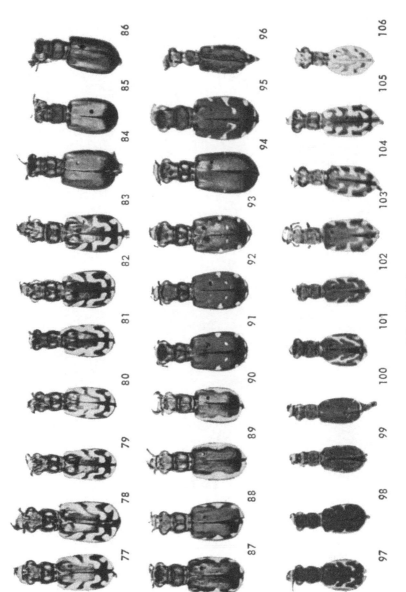

PLATE 4

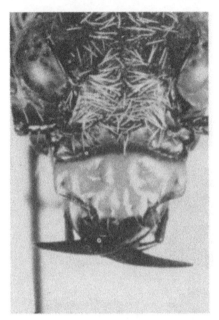

FIG. 1

FIG. 3

FIG. 2

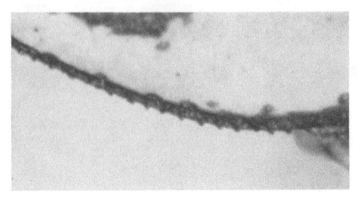

FIG. 4

are darker in colour, with complete humeral and apical lunules, and with the middle band usually expanded briefly along the margin (No. 41, Plate 2). However, in the more northern populations, this phenotype occurs sparingly, and it is connected by a series of intergrading phenotypes, individuals having the reduced markings. The variation may be clinal, but the details have not been worked out.

SYNONYMY

Typical *duodecimguttata* is illustrated in No. 40, Plate 2. Casey's *hudsonica* (Nos. 42–44, Plate 2) is a variant with the elytral markings further reduced. The name *bucolica* Casey (No. 41, Plate 2) has been applied to the phenotype which is dominant in the southern portions of the prairie provinces. For further remarks relating to this latter form, see the discussion of synonymy under *repanda*.

RELATIONSHIPS

This species is very similar to *repanda*, and so the two may be closely related. A species even more similar is *oregona* Le Conte. The latter and *duodecimguttata* share a similar pattern of elytral markings, and both have identical male genitalia. In Canada, at least, the two species are allopatric. Perhaps they are really only well-marked subspecies.

HABITAT

The requirements of this species are similar to those of *repanda*. For details, see the discussion of habitat of the latter species.

LOCALITIES

Map 1 indicates the positions of the localities which are listed here: NEWFOUNDLAND, Badger, Buchan's Junction, Cape Ray, Codroy, Little River Codroy, Corner Brook, Deer Lake, South Brook, Gambo, Gander, Garnish, Glenbournie, Glenwood, Grand Bank, Great Rattling Brook, Hampden, Harmon Field, Humber River, King's Point, Lomond, Millertown Junction, Nicholsville, Piccadilly, Port-aux-Basques, St. Fintans, Shoal Harbour, South Branch, Springdale, Steady

FIG. 1. Head of *Cicindela formosa manitoba*, illustrating the character "head hairy."

FIG. 2. Head of *Cicindela oregona*, illustrating the character "head with only clusters of hairs." Note also the short labrum.

FIG. 3. Head of *Cicindela longilabris*, illustrating the character "head glabrous." Note also the long labrum of this species.

FIG. 4. Serrulations along the apex of an elytron.

Brook, Stephenville, Victoria Lake; NOVA SCOTIA (including Cape Breton), Baddeck, Barrington Passage, Dartmouth, Halifax, Ingonish, Cape Kentville, Yarmouth; NEW BRUNSWICK, Bathurst, Shediac; QUEBEC, Cascapedia Lake (Mount Lyall), Kingsmere, Knowlton, Natashquan, Otter Lake (40 miles west of Ottawa), South Bolton; ONTARIO, Favourable Lake, Lat. 53° N., Glen Williams, Kapuskasing, Lobo, Muskoka, Ogoki, Ottawa, Sault Ste Marie, Ventnor, Woodstock; MANITOBA, Aweme, Baldur, Berens River, Birds Hill, Delta, Glenboro, Husavick, The Pas, Mile 256 Hudson Bay Railway, Riding Mountain, Roland, Rounthwaite, Shoal Lake, Victoria Beach, Westbourne; SASKATCHEWAN, Christopher Lake, Estevan, Fish Lake, Good Spirit Lake, Holbein, Maryfield, Pike Lake Reserve, Saskatoon, Torch River, Waskesiu; ALBERTA, Bilby, Cooking Lake, Edmonton, Happy Valley, Lesser Slave Lake, Medicine Hat, Tofield; NORTHWEST TERRITORIES, Fort Smith; BRITISH COLUMBIA, Vancouver (near campus, University of British Columbia). Most if not all early British Columbia records should be referred to *oregona*.

Cicindela oregona Le Conte, 1857
(Nos. 70–75, Plate 3, and Map 2)

This species may be distinguished at once from all other tiger beetles, except females of *scutellaris*, by the confinement of the vestiture of the head to a small subcircular patch of long hairs near the inner front edge of each eye. These hairs may be abraded; if so, their former position is indicated by a roughened and punctate area. The markings of the elytra are basically similar to those of typical *duodecimguttata*, but they are subject to more extensive disintegration.

GEOGRAPHICAL VARIATION AND SUBSPECIES

This species ranges widely in the west, extending from southern New Mexico northward to southern Alaska (Fairbanks). Only the Canadian populations will be described here. Four characters vary geographically: elytral markings, lustre of dorsal surface, colour of thoracic pleura, and form of elytral apices.

Populations which inhabit the eastern slopes of the Rocky Mountains are marked as in No. 73, Plate 3, with the humeral lunule represented by two large dots. The dorsal surface is more or less olive, and has a pronounced metallic lustre. The ventral surface is bicoloured, with the thoracic pleura coppery (No. 74, Plate 3). The elytral spine is small, and the serrations at the apex are feeble. The name *oregona guttifera* Le Conte applies to specimens with these characteristics.

The southern half of British Columbia, excluding the Cascades and

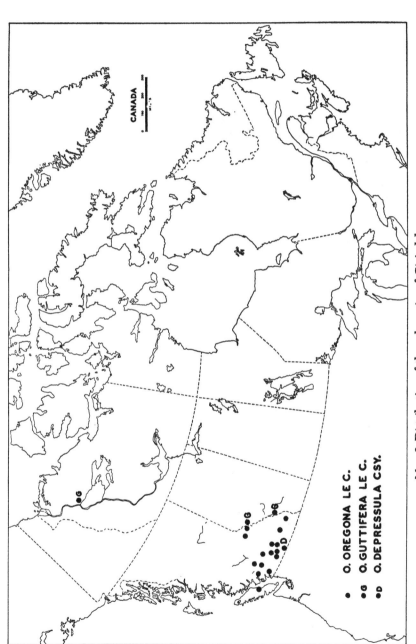

● O. OREGONA LE C.

●G O. GUTTIFERA LE C.

●D O. DEPRESSULA CSY.

MAP 2. Distribution of the subspecies of *Cicindela oregona*

Vancouver Island, is occupied by populations with markings and colours of the dorsal surface like those of *o. guttifera*, but with the elytral spine and serrulations well developed, and the thoracic pleura metallic blue, like the rest of the ventral surface (Nos. 70–71, Plate 3). These features characterize the nominate subspecies, *oregona oregona*.

In the Cascades there are occasional specimens which are like *oregona oregona* except that their elytral markings are somewhat reduced and their dorsal surface bright blue (No. 72, Plate 3). The name *scapularis* Casey has been applied to them, but I regard them as *oregona oregona*. For a discussion of their status, see below, under Synonymy.

Vancouver Island and the Cascades are occupied by a third group of populations, called *oregona depressula* Casey. Specimens of this subspecies have the elytral markings reduced, the humeral lunule frequently absent, the integument of the dorsal surface more pronouncedly metallic green, the thoracic pleura coppery or metallic green (not blue), and the elytral serrulations and spine reduced (No. 75, Plate 3).

Synonymy

The name *Cicindela audax* Casey has been applied to Canadian specimens which are here referred to *guttifera*. However, study of the original description of *audax* indicates that this name refers only to variant individuals within populations the majority of which are typical *o. guttifera*. Thus *audax* is a junior synonym of *guttifera*.

Perhaps the form *scapularis* Casey (described above) should be accorded subspecific status, but I choose to regard it as *oregona oregona* pending investigation of more extensive material from critical areas.

Relationships

See the discussion of relationships under *duodecimguttata*.

Habitat

Cicindela oregona lives along the margins of lakes and streams, on clay or sandy soil. It seems to occupy about the same niche as *repanda*, for the two species occur together commonly over much of southern British Columbia.

Localities

Map 2 indicates the positions of the localities which are listed below.

oregona guttifera Le Conte: NORTHWEST TERRITORIES, Good Hope; ALBERTA, Athabasca Falls, Banff

oregona oregona Le Conte: ALBERTA, Banff; BRITISH COLUMBIA, Cariboo Road (Mile 185), Comox, Fort Steele, Coldstream, Hedley, Kamloops, Kaslo, Lillooet, Mabel Lake, Merritt, North Bend, North Vancouver, Oliver, Pender Harbour, Powell River, Sooke, Sugar Lake, Vaseaux Lake, Vernon

oregona depressula Casey: BRITISH COLUMBIA, Tod Inlet, Vancouver Island; Cascades (Hatch)

Cicindela hirticollis Say, 1817
(Nos. 49–50, Plate 3, and Map 3)

The only species in our area with which *hirticollis* is likely to be confused are *nevadica* and *cuprascens*, from which it may be at once separated by the more or less erect hairs above and below, these being decumbent in the other two.

GEOGRAPHICAL VARIATION

This exceptionally widespread species is found southward around the coast of North America from Labrador to British Columbia, and is widely distributed inland. There is some difference in the width of the markings, but the Canadian populations on the whole are quite widely marked, the marginal band attaining the humeral lunule and almost or quite reaching the apical lunule (Nos. 49–50, Plate 3). Most individuals are large enough and with markings sufficiently broad to be called *h. ponderosa*, if that form is worth naming, but slightly less brilliant. A few Manitoba specimens have the markings considerably reduced, but they were taken in the same locality as more typical specimens.

RELATIONSHIPS

Hirticollis resembles *repanda* more closely than any other species and may therefore be closely related to it.

HABITAT

This species shows a marked preference for dry, light-coloured sand, particularly whitish, and where such conditions obtain it frequently occurs in large numbers.

LOCALITIES

Map 3 indicates the positions of the localities which are listed here:

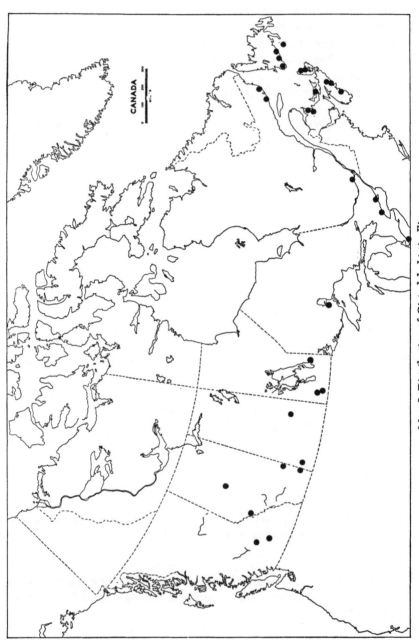

MAP 3. Distribution of *Cicindela hirticollis*

MIQUELON ISLAND (French possession); NEWFOUNDLAND, Burgeo, Grand Bruit, Port-aux-Basques; NOVA SCOTIA, Summerville, White Point Beach (Queens County), Cow Bay (Halifax County), Ingonish (Cape Breton); PRINCE EDWARD ISLAND, Deroche Point; NEW BRUNSWICK, Tracadie, Tabusintac; QUEBEC, Lachute, Natashquan, Thunder River; ONTARIO, Hastings County, Lake Nipigon, Metanick River, Point Pelee, Toronto; MANITOBA, Aweme (16 miles east), Glenboro, Victoria Beach; SASKATCHEWAN, Good Spirit Lake, Hatton, Rudy, Yonker; ALBERTA, Empress, Jasper Lake, Lesser Slave Lake, Medicine Hat (as *ponderosa*); BRITISH COLUMBIA, Cariboo Road (Mile 185), Kamloops, Morley, Wasa Lake near Kimberley

Cicindela limbata Say, 1823
(Nos. 22–26, Plate 2, Map 4)

The southern populations of this species, comprising individuals with predominantly whitish elytra, differ markedly from all other Canadian *Cicindela*, except *formosa gibsoni*. However, *limbata* is a much smaller species than *formosa*, so it is easy enough to distinguish between the two (cf. No. 23, Plate 2, and No. 14, Plate 1). Specimens of *limbata* from farther north have an elytral pattern similar to that of *repanda*, but in the former, the lower arm of the humeral lunule is straight, whereas in *repanda* this line is curved (cf. No. 26, Plate 2, and No. 45, Plate 2).

GEOGRAPHICAL VARIATION AND SUBSPECIES

Marked variation in this species occurs in extent of colour markings, although the colour itself exhibits slight variation. Three groups of populations, two of which are found in Canada, can be distinguished on the basis of these features.

Specimens which inhabit the prairies and southern reaches of the boreal forest have elytra which are predominantly yellowish-white; a narrow strip bordering the suture is brown-bronze (No. 23, Plate 2). A small percentage of specimens have brown-bronze subapical spots as well (No. 24, Plate 2). The name *limbata nympha* Casey applies to these populations. In the northern portions of the United States, *limbata limbata* is found (No. 22, Plate 2). In this subspecies the dark areas of the elytra are green-bronze and the subapical spots are usually present; these colour markings distinguish it from *limbata nympha*.

A few specimens with the characteristics of the nominate subspecies, *l. limbata*, have been found within the range of *l. nympha*. For example, during 1954, A. R. Brooks and I collected nearly 200 specimens of *C. limbata* at six localities in the prairie of Saskatchewan, and at one

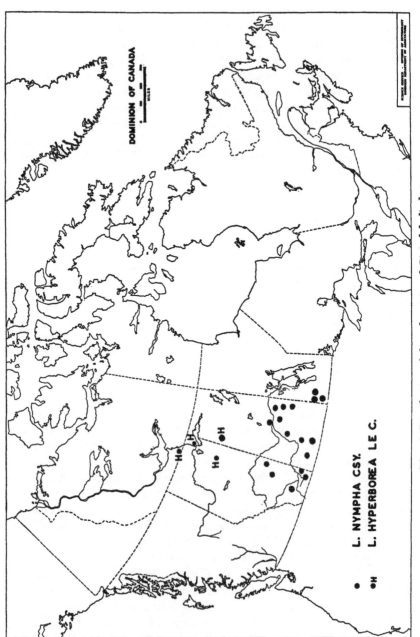

MAP 4. Distribution of the subspecies of *Cicindela limbata*

locality in Manitoba. In only two specimens were the dark areas of the elytra green-bronze, rather than brown-bronze, but the specimens did not otherwise approach *l. limbata*. In a third specimen the elytra had more extensive dark areas than usual. The remaining specimens were typical *l. nympha*.

The third group of populations includes specimens which exhibit a relatively pronounced increase in dark areas on the elytra (Nos. 25–26, Plate 2). These are called *limbata hyperborea* Le Conte, and the types of this subspecies were collected at the Methy Portage, 57° N. latitude, in Saskatchewan.

Evidence of an extensive blend zone between *l. nympha* and *l. hyperborea* has not been found. Populations of *l. nympha* and a few specimens with characteristics of *l. hyperborea* have been taken at Edmonton. At Chauvin, Alberta, which is in the parkland south and east of Edmonton, one specimen was taken with pigmentation approaching that of *l. hyperborea*. Other specimens taken in the same locality were typical *l. nympha*.

Farther north, at McMurray, Alberta, and at Fort Smith, Northwest Territories, only typical *l. hyperborea* has been found. It seems virtually certain that blend populations between these two subspecies exist, and they should be sought in the areas north of the latitude of Edmonton.

Synonymy

The name *limbigera* Gemminger and Harold, 1868, was proposed for this species. W. Horn (1915), however, retained the name used here.

Mrs. Vaurie (1950) recorded *limbata limbata* from Canadian localities. However, she did not recognize *nympha* as a distinct subspecies, and her records almost certainly refer to the latter form.

Relationships

This species is closer to the preceding ones than to any other North American tiger beetles. However, the form of the male genitalia and of the elytral markings sets it clearly apart even from those which it resembles most strongly.

Habitat

Sandy areas, usually remote from water, are the sites favoured by this species. In the boreal forest, it occurs in blow-outs along jack pine ridges, in the company of *tranquebarica* and *lengi*.

LOCALITIES

Map 4 indicates the positions of the localities which are listed below.

limbata nympha Casey: MANITOBA, Aweme (type locality), Deleau, Glen Souris, Oak Lake; SASKATCHEWAN, Beaver Creek, 9 miles northeast of Canora, south of Caron, Elbow (Qu'Appelle Valley), Gascoigne, Good Spirit Lake, Great Sand Hills, Holbein, Lake Chaplin, Lake Katepwa, Prince Albert, Pike Lake, Radisson, Rudy, Sceptre, southern Saskatchewan (Vaurie, 1950), Torch River, Yonker; ALBERTA, Chauvin, Claysmore, Czar, Edgerton, Empress, Medicine Hat, Orion, Ribstone, southern Alberta (Vaurie, 1950)

limbata hyperborea Le Conte: MANITOBA, Cormorant Lake (very doubtful); SASKATCHEWAN, Methy Portage; ALBERTA, Edmonton, Fort Chipewyan, Fort McMurray; NORTHWEST TERRITORIES, Fort Smith (60° N.)

Cicindela sexguttata Fabricius, 1775
(Nos. 91–94, Plate 4, and Map 5)

This species, with its brilliant green integument, and either with or without six to ten white dots on the elytra, is not likely to be confused with any other tiger beetle in eastern North America. Some specimens, however, have characteristics which approach those of *patruela*.

GEOGRAPHICAL VARIATION AND SUBSPECIES

Colour, lustre, and maculation of the elytra vary geographically. Over much of the range of *sexguttata*, populations are typically maculate (No. 91, Plate 4), usually with eight white dots, and the colour of the integument is green, blue-green, or violaceous. Such populations represent the subspecies *sexguttata sexguttata*. Occasionally in populations comprising mainly maculate specimens, a few immaculate individuals are found.

In western Ontario and the adjacent parts of Manitoba, the subspecies *sexguttata denikei* occurs. This form is usually immaculate, and only rarely has more than inconspicuous median, submarginal, and apical spots. Its colour is olivaceous green, occasionally showing a bluish reflection (No. 94, Plate 4). This form is larger on the average, and is a little flatter than the nominate subspecies. The most convincing evidence for according taxonomic recognition to *denikei* is derived from its ecology. For details, see below.

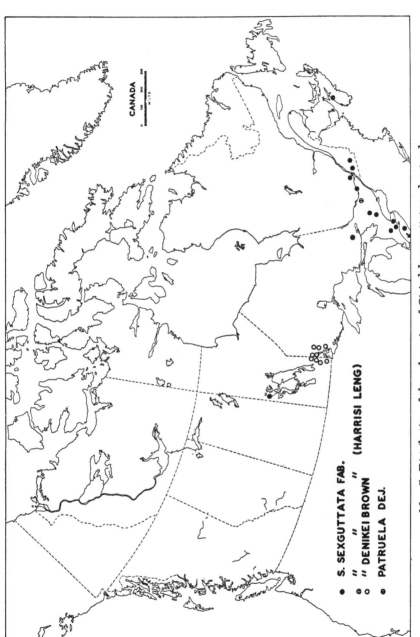

CANADA

0 100 200 300
scale

● S. SEXGUTTATA FAB.
⊖ " " DENIKEI BROWN (HARRISI LENG)
○ " "
◐ PATRUELA DEJ.

Map 5. Distribution of the subspecies of *Cicindela sexguttata* and *patruela*

SYNONYMY

The name *harrisi* Leng has been used for the immaculate, shiny specimens taken with typical *s. sexguttata* (No. 93, Plate 4). This form seems to be only an individual variant, and so should not be recognized taxonomically. I regard it as a synonym of *s. sexguttata.*

RELATIONSHIPS

The marked similarity in structure of the male genitalia as well as in the colour and markings of the elytra exhibited by *sexguttata* and *patruela* suggest that the two species are closely related. However, it is not possible to relate them to any other North American *Cicindela.*

HABITAT

The preferred habitat of *s. sexguttata* appears to be shady paths through rich, deciduous woods, or in pastures near such situations. This subspecies is abundant in Ontario and Quebec. The atypical sub-species, *sexguttata denikei*, is found along the edge of the Laurentian highland. I have myself taken it on the summit of rocky knolls in the middle of heavy stands of conifers. In such situations it runs rapidly over the hot, bare rocks, and when pursued, instead of flying, will take refuge under a little oasis of green moss. In spring it is found on sandy or gravelly soil, in the open, but always close to spruce or pine. Since *s. sexguttata* and *s. denikei* differ so much in their habitats, it is possible that the two are specifically distinct. If so, then they are sibling species because they are so similar in morphological characters, including the structure of the male genitalia.

LOCALITIES

Map 5 indicates the positions of the localities which are listed below. One questionable locality record for *s. denikei* has been noted. A dilapidated specimen, said to have been taken at Morden, Manitoba, was included in a student collection received at the University of Manitoba. It is certainly this subspecies, but as Morden is far from the coniferous zone, the record awaits confirmation.

sexguttata sexguttata Fabricius: NOVA SCOTIA, Brickton, Kentville (immaculate); QUEBEC, Aylmer, Chelsea, Covey Hill (maculate and immaculate), Fairy Lake, Ile Jésus, Knowlton (maculate and immacu-late), Montreal (maculate and immaculate), Old Chelsea, Paugan (maculate); ONTARIO, Bell's Corners near Ottawa, Ottawa (both immaculate), Constance Bay, Fenelon Falls, Fisher Glen, Hymers, Jack River, Leamington, Miner's Bay, Niagara Glen, North Hatley,

Ottawa, Sudbury, Turkey Point, Walsingham (all maculate); MANITOBA, Makinak (almost certainly an error for *s. denikei*)

sexguttata denikei Brown: ONTARIO, Ingolf (type locality, just inside Ontario boundary; most of the type series actually taken in Manitoba), Kenora, Malachi, Sioux Narrows; MANITOBA, Altona, Betula Lake, Lake Brereton, Sandilands Forest Reserve, Slave Falls, Whiteshell Forest Reserve

Cicindela patruela Dejean, 1825
(No. 95, Plate 4, and Map 5)

This species most closely resembles *sexguttata sexguttata*. However, the integument of *patruela* is duller, and the middle band of the elytron is complete.

RELATIONSHIPS

See discussion of this topic under *sexguttata*.

HABITAT

So far as is known, *patruela* is found in Canada at but one station. Of this, W. J. Brown writes: "There is a small colony, never too populous, at Constance Bay, which is on the Ottawa River, about 24 miles west of Ottawa. The area is sand and supports a growth of jack pine (the most abundant tree), sweet fern and blueberry. The colony seems to be concentrated largely on one small stretch of sandy lane." Brown also notes that "*C. sexguttata* . . . occurs at Constance Bay, but I have never seen it flying on that lane with *patruela*."

Cicindela scutellaris Say, 1823
(Nos. 84–90, Plate 4, and Map 6)

This species is well characterized by the white labrum and hairy front of the male, and, in the female, by the black labrum and the loose cluster of a few hairs near the front inner edge of each eye. Females of *oregona* also have the vestiture of the head confined near the front inner edge of each eye.

GEOGRAPHICAL VARIATION AND SUBSPECIES

No other species of North American tiger beetles exhibits the range of variation in colour and markings seen in *scutellaris*. In Canada, three forms occur. From Ontario and Quebec in the east to southeastern Manitoba in the west, the elytra may be dull in colour or bright green, olive green, dull red, or bright red. The elytra are maculate laterally,

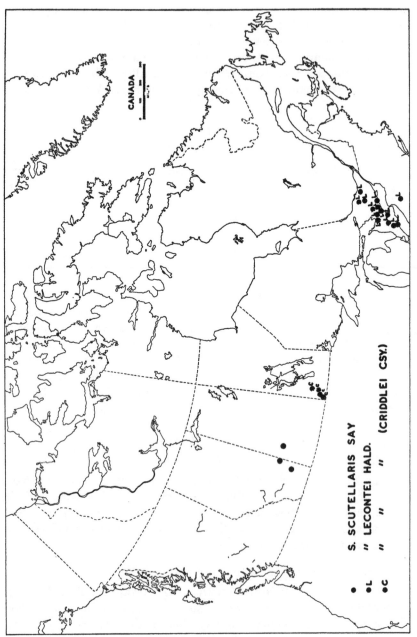

MAP 6. Distribution of the subspecies of *Cicindela scutellaris*

CANADA

● S. SCUTELLARIS SAY
●L " " LECONTEI HALD.
●C " " (CRIDDLEI CSY.)

and these maculations may form a broken band with irregular edges (Nos. 87–88, Plate 4) or a complete band with less jagged edges (Nos. 89–90, Plate 4). In specimens from the more eastern localities the dominant condition is a broken, narrower band, whereas in the samples from Manitoba it is a broad, unbroken band. Probably the variation is clinal, and the subspecific name *scutellaris lecontei* Haldeman may be applied to the entire complex. For those who wish to have a formal designation for the more western specimens with the wider elytral band, the name *scutellaris criddlei* Casey may be used (No. 90, Plate 4; type locality Aweme, Manitoba).

In southwestern Saskatchewan and southeastern Alberta, this species is represented by immaculate populations with the dorsal integument of the head and pronotum brilliantly metallic blue or green and the elytra equally brilliant—coppery or green or any mixture of the two colours. This is the nominate subspecies, *scutellaris scutellaris* Say (Nos. 84–86, Plate 4).

SYNONYMY

The name *criddlei* Casey seems to apply to the terminal components of a cline in the development of the lateral elytral band. This name is best regarded as a synonym of *scutellaris lecontei*.

RELATIONSHIPS

This species does not appear to have any close extant relatives in the North American tiger beetle fauna. Its genitalia are more like those of *formosa* than like those of any other species, but the resemblance is not striking.

HABITAT

In Canada, this species is frequently found in the company of *formosa*. It seems to prefer sandy blow-outs, where the sand is loose and the vegetation very sparse. It does not go far north, even though the sandy jack pine ridges of the boreal forest would seem to offer suitable habitats.

LOCALITIES

Map 6 indicates the positions of the localities which are listed below.

scutellaris scutellaris Say: SASKATCHEWAN, Great Sand Hills; ALBERTA, Empress, Medicine Hat

scutellaris lecontei Haldeman: QUEBEC, Guyon; ONTARIO, Britannia, Constance Bay, Hamilton, Hyde Park Corner, Kendal, Kettleby, Leskard, London, Merivale, Muskoka, Paris, Point Pelee, Tillsonburg,

Toronto, Wasaga, Willowdale; MANITOBA, Aweme, Hartney, Melita, Oak Lake

Cicindela formosa Say, 1823
(Nos. 3–16, Plate 1, and Map 7)

The nominate subspecies, *formosa formosa*, may be mistaken for *lengi*, but the characters presented in the key enable one to distinguish between these two superficially very similar, but structurally very different, species. Otherwise, this species is easily recognized by its large size, and the form of the humeral lunule, if it is present.

GEOGRAPHICAL VARIATION AND SUBSPECIES

The following characters exhibit variation: lustre of the ventral surface, and colour and extent of the pigmented areas of the dorsal surface. Four subspecies are recognized, based on variation in these characters.

Specimens from eastern United States and Canada, which are referred to the subspecies *formosa generosa* Dejean, are dull, not metallic, on the underside; the pigmented areas of the elytra are blackish-red, brown, or green, and they contribute to the formation of a complete dorsal pattern (No. 9, Plate 1). In two out of thirty-three specimens before me, the pigmented areas are reduced (No. 10, Plate 1). In four specimens the pigmented areas are red-brown, and in three of these the ventral surface is metallic, whereas in the fourth it is dull, as is usual for this subspecies.

Specimens from Aweme (109 individuals) and Oak Lake, Manitoba (twenty-four individuals), and Denbigh, North Dakota (thirty-five individuals), exhibit bright metallic undersides, with coppery reflections on the sterna. The pigmented areas of the dorsum are red-brown, usually with a feeble purplish reflection. The white areas, on the average, are considerably greater in extent than in *f. generosa*, and this difference is shown especially well by the much shorter humeral lunules (cf. Nos. 4–8, Plate 1, and Nos. 9–10, Plate 1). An extreme of variation is exhibited by a specimen collected at Denbigh, North Dakota, in which the pigmented areas are reduced to such an extent that the humeral lunule is completely absent, and so is the post-median bar. The Manitoba and North Dakota populations described here represent the subspecies *formosa manitoba* Leng, the type locality of which is Aweme, Manitoba.[1]

[1]Aweme, Manitoba, is situated at the southwest edge of a sand hill area extending from the Assiniboine River west of Shilo eastward well towards Portage la Prairie. Excluding small extensions north and south, the maximum width of the area is about 16 miles.

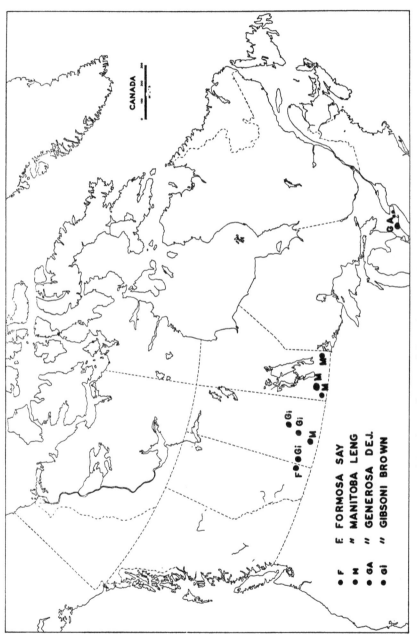

Map 7. Distribution of the subspecies of *Cicindela formosa*

A third subspecies, *formosa gibsoni* Brown, is known from the follow-ing localities in Saskatchewan: Great Sand Hills (type locality, type series, fifteen specimens); Elbow (twenty-two specimens); Pike Lake (sixty-eight specimens). This subspecies is also known from Maybell, Colorado, several hundred miles south of the Saskatchewan localities. It is surprising to find that the range of *formosa formosa* extends between the Colorado and Saskatchewan pockets of *gibsoni*. In *f. gibsoni*, the underside of the body is metallic, but lacks the coppery reflections seen in *f. manitoba*. The pigmented areas of the dorsum are darker, more purplish than in *f. manitoba*. The most striking character-istic of this subspecies, however, is the great reduction in pigmented areas of the elytra. Over one-half of the specimens examined exhibit the pattern illustrated in No. 13, Plate 1. Sixteen specimens from Pike Lake have the pigmented areas further reduced (No. 14, Plate 1). The remaining specimens have a more complete colour pattern and about 3.8 per cent of the total number of specimens are within the range of variation of topotypical *f. manitoba*. With the exception of the extreme variants of the Pike Lake population sample, the remaining phenotypes are of about the same frequency in each of the population samples listed above. Thus they are all about equally different from the other subspecies of *formosa*.

The fourth subspecies, *formosa formosa*, ranges throughout the western portion of the Mississippi Basin. It is known from only a few Canadian localities, all in Alberta. The ventral surface in this sub-species is bright metallic, the pigmented areas of the dorsum form a complete colour pattern, and are brilliant red-purple (No. 3, Plate 1). It is interesting to note that the distance separating the colony of *formosa formosa* in Empress, Alberta, from the colony of *f. gibsoni* in the Great Sand Hills is only 50 miles. In spite of this, intergrading specimens have not been found at either site.

SYNONYMY

I have examined two paratopotypes of *formosa fletcheri* Criddle, 1925 (type locality Marias River, Montana), and there is no doubt that this named form is within the range of variation of *f. formosa*. As W. Horn indicated, *fletcheri* should be listed as a synonym of *f. formosa*.

RELATIONSHIPS

This species seems to be without close extant relatives, but see remarks under *scutellaris*.

HABITAT

This species dwells in sandy areas, but there is some difference in the choice of specific habitats among the subspecies. The nominate subspecies and *f. generosa* are satisfied with areas of yellowish sand that have bare spots and sparse vegetation. Both *f. manitoba* and *f. gibsoni* favour blow-outs and drifting sand areas on the edges of which they can get the protection of encroaching vegetation.

In southern Saskatchewan and Alberta, there are a number of sandy areas which appear to be suitable for *formosa*, but the species is not present. These same areas are occupied by the species *lengi* and *limbata*, and these species are taken elsewhere with *formosa*. These facts suggest that, although the sites which *formosa* has not occupied appear suitable, they really do not fulfil its requirements. This problem should be investigated.

LOCALITIES

On Map 7 are indicated the positions of the localities listed below.

formosa generosa Dejean: ONTARIO, Fisher Glen, Normandale, Walsingham

formosa manitoba Leng: MANITOBA, Aweme (Spruce Woods Forest Reserve), Douglas Lake (Spruce Woods Forest Reserve), Glenboro, Grunthal, Kelwood (most probably incorrect), Oak Lake, Onah (Spruce Woods Forest Reserve), Pipestone; SASKATCHEWAN, Shaunavon

formosa gibsoni Brown: SASKATCHEWAN, Beaver Creek, Elbow (Qu'Appelle Valley), Fox Valley (Great Sand Hills), Pike Lake, southeast of Elbow

formosa formosa Say: ALBERTA, Empress, forks of Saskatchewan River near Red Deer River, Fort MacLeod, Medicine Hat

Cicindela cimarrona Le Conte, 1868
(No. 37, Plate 2)

The maculation of the elytra of this species is sufficient to distinguish it from all other Canadian *Cicindela* except fully marked *longilabris* and *montana* (No. 65, Plate 3). However, the labrum of *cimarrona* is much shorter than the labra of the other two species.

RELATIONSHIPS

This species and *decemnotata, limbalis,* and *purpurea* seem to form a natural unit, for they are similar in external characteristics, and the males of these species have identical genitalia. *Cimarrona* has the most complete markings and may be the most primitive species of these four. This group of species does not seem to be closely related to any other extant North American *Cicindela*.

HABITAT

Bare clay is said to be the soil preferred by *cimarrona*.

LOCALITY

British Columbia, Canal Flats (G. Stace-Smith, collector), Victoria (Vancouver Island). I have been unable to verify this latter record.

Cicindela decemnotata Say, 1817
(No. 38, Plate 2, and Map 8)

The green or violaceous integument of the dorsal surface, the reduced humeral lunules, and the relatively long descending arms of the middle bands characterize this species.

VARIATION

The dorsal integument of specimens from the prairie provinces is predominantly green, but in specimens from Whitehorse, Yukon Territory, the head and pronotum are coppery-violaceous, and the elytra are violaceous.

SYNONYMY

The species *albertina* Casey, which is very similar to *decemnotata*, was described as having the "apical lunule . . . strongly divided." However, this is a characteristic not possessed even by the majority of Alberta specimens, and it was from this province that the species was described. I regard this name as a synonym of *decemnotata*.

RELATIONSHIPS

See the remarks under *cimarrona* Le Conte.

HABITAT

This species occurs in Canada only in the west. It has been collected on dry, gravelly clay soil.

LOCALITIES

The positions of the localities which are listed here are indicated on Map 8: SASKATCHEWAN, Eastend, Elbow, Swift Current; ALBERTA, Dorothy, Edmonton, Lethbridge, Fort MacLeod, Medicine Hat; BRITISH COLUMBIA, "south and southwest"; YUKON TERRITORY, Whitehorse

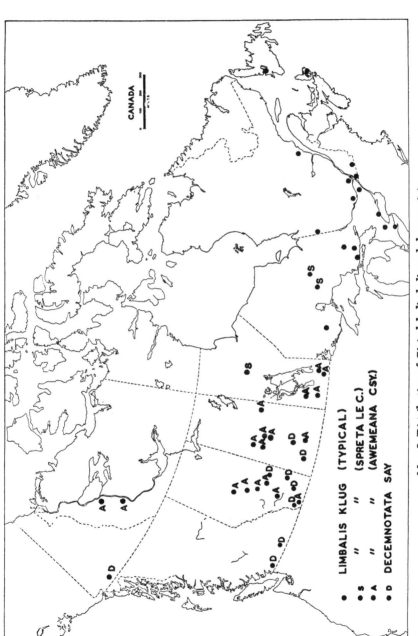

●   LIMBALIS KLUG (TYPICAL)

●S   "   "   (SPRETA LE C.)

●A   "   "   (AWEMEANA CSY.)

●D   DECEMNOTATA SAY

MAP 8. Distribution of *Cicindela limbalis* and *decemnotata*

Cicindela limbalis Klug, 1834
(Nos. 34–36, Plate 2, and Map 8)

The characters given in the key are sufficient to distinguish this species from all others found in Canada.

GEOGRAPHICAL VARIATION

The colour of the dorsal integument is rather dull red in individuals from localities in central and eastern Canada (No. 34, Plate 2). In Manitoba, and farther west, specimens are more brilliantly coloured (No. 35, Plate 2). Farther north, populations are found in which the integument of the dorsum is suffused with green (No. 36, Plate 2).

SYNONYMY

The central and eastern populations are typical *limbalis*. The name *awemeana* Casey (type locality, Aweme, Manitoba) was proposed for the more brilliantly coloured specimens found in western Canada. The name *spreta* Le Conte applies to the more greenish, northern populations. In my opinion, the differences between these forms are too slight to justify subspecific status. Therefore, all of the Canadian populations may be included under the name *limbalis* Klug, of which *awemeana* Casey and *spreta* Le Conte are synonyms.

RELATIONSHIPS

See the remarks under *cimarrona* Le Conte. This species is certainly closer to *purpurea* in its external characteristics than to the other two species of this complex. Although the morphological differences between *purpurea* and *limbalis* are slight, the ecological differences which they exhibit are well marked, and as the gross ranges of the two overlap extensively, they must be regarded as distinct species.

HABITAT

In my experience, *limbalis* prefers banks of heavy clay, which are quite steep and bare. These observations bear out the results of Shelford (1908). He was able to show that this species, when given the choice of soil types and between steep and level ground produced forty-six larvae to one in favour of clay and forty-six to one in favour of a steep incline. One larva appeared in soil consisting of clay mixed with humus, and no larvae were produced from pure sand.

LOCALITIES

On Map 8 are indicated the positions of the localities listed here:

NEWFOUNDLAND, Bay of Islands; NOVA SCOTIA, Cape Breton; QUEBEC, Covey Hill, Kazebazua, Knowlton, Old Chelsea, Opasatika Lake, Three Rivers; ONTARIO, Algoma, Guelph, Nakina, Newcastle, Normandale, Ottawa, Sault Ste Marie, Sioux Lookout, Smoky Falls on Mattagami River, Sudbury, Toronto; MANITOBA, Aweme, Mile 332 Hudson Bay Railway, Oak Lake, Riding Mountain, Roland, Shell River, Treesbank, Wawanesa, Winnipeg; SASKATCHEWAN, Asquith, Cut Knife, Moose Jaw, North Battleford, Saskatoon, Swift Current, Toch River north of White Fox; ALBERTA, Bilby, Calgary, Edmonton, Fawcett, Happy Valley, Lausand, Nevis, Pincher Creek, Wabamun, Wetaskiwin; NORTHWEST TERRITORIES, Fort Norman, Fort Wrigley

Cicindela purpurea Olivier, 1790
(Nos. 27–33, Plate 2, and Map 9)

The majority of specimens of this species lack a subhumeral dot on the elytra, and thus may be readily distinguished from *limbalis*, the species to which they are most similar. The more oblique shorter middle band is also a useful recognition character. Very occasionally among a population of *purpurea*, individuals may be found in which a subhumeral dot is present, or in which the middle band approaches quite close to the side margin. Individuals such as these, when of the colour of *limbalis*, might be mistaken for that species. However, I have seen none with the colour of *limbalis* in which the more oblique, shorter middle band did not show its relationship to *purpurea*.

GEOGRAPHICAL VARIATION AND SUBSPECIES

In Canada, three groups of populations may be discerned, based on average differences in colour. In eastern specimens the pronotum is bronzed, and the elytra are cupreous with green margins and suture. Populations of such individuals represent the nominate subspecies, *purpurea purpurea* (No. 27, Plate 2). In the prairie provinces, the elytra vary through all shades of green or blue-purple to pure green, or black, with a more or less metallic sheen. Specimens that have black elytra also have black pronota and represent a melanic colour phase. In other specimens the pronotum is bronzed, as in the eastern populations, or rarely green, as in individuals from British Columbia. This group of populations is called here *p. purpurea* x *p. auduboni* (Nos. 28–31, Plate 2). The name *auduboni* was proposed for a specimen with pure green elytra from farther south. The majority of Colorado specimens which are referred to *p. auduboni* have either pure green elytra or at least approach that colour. Possibly the Canadian popula-

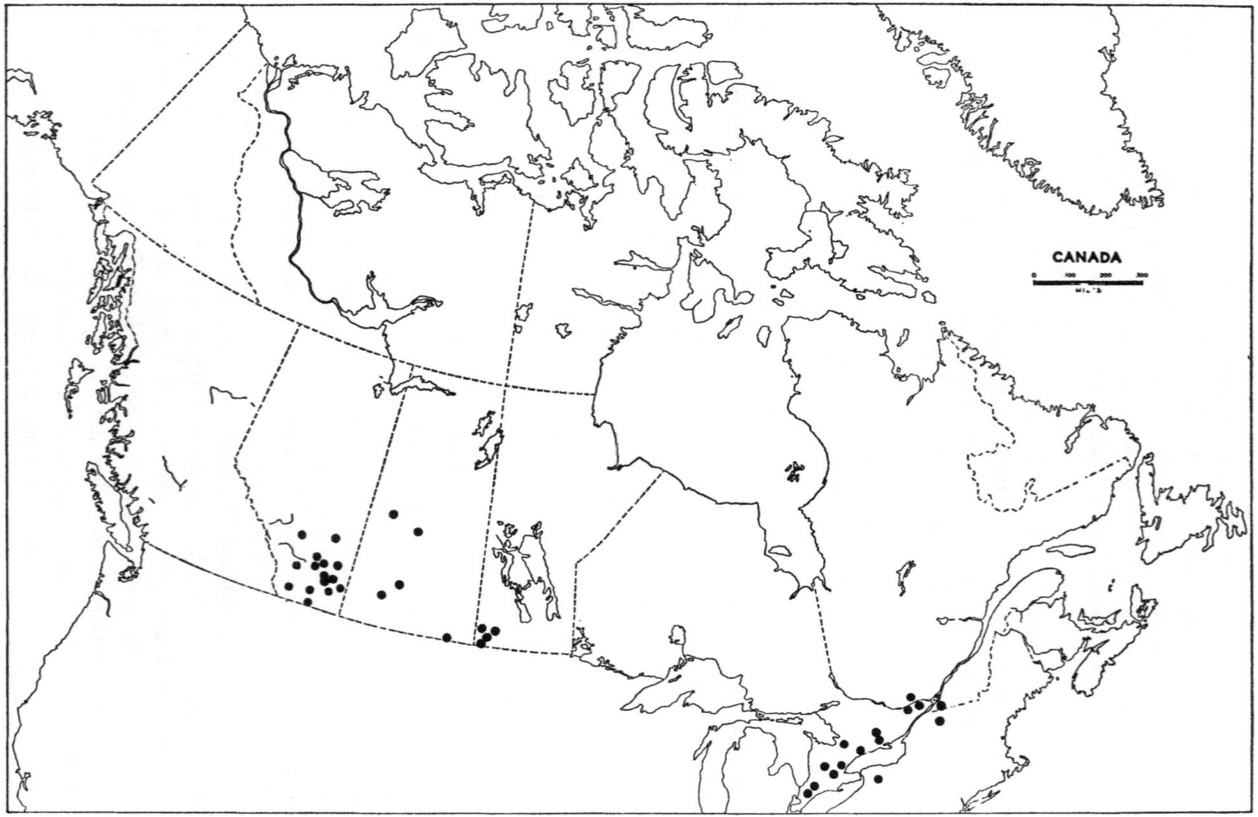

MAP 9. Distribution of *Cicindela purpurea*

tions form part of a cline from nominate *purpurea* of the east to the *p. auduboni* of the west.

Two distinct colour phases are found in British Columbia populations of *purpurea*: a melanic one, and a "normal" one. The latter has a pure green pronotum and pure green elytra, and has received the name *purpurea pugetana* Casey (No. 33, Plate 2). This name must be applied to both colour phases.

The occurrence of melanism in this species is interesting. Smyth (1935) claims that the dark form which occurs in the Canadian prairies (*nigerrima* Leng, No. 32, Plate 2) is of different genetic composition and of different habits than the non-melanic *purpurea*. That the two are different genetically is highly probable, but both seem to have the same habits, Smyth's statement notwithstanding. Melanics and non-melanics are taken together, and cross-matings are seen commonly.

Typical markings of *purpurea* consist of an oblique middle band which does not attain the margin, and an apical dot. These markings, however, are subject to extensive variation. For example, in a series of over 100 specimens from Brandon, Manitoba, variation extends from almost immaculate specimens to those having a humeral dot, a middle band reaching the margin and extending a short distance along it, and ante-apical and apical dots. The majority of melanic specimens exhibit typical *purpurea*-type markings, but a few immaculate individuals have been seen.

This extensive variation in colour and the geographical pattern which it portrays require detailed analysis from the standpoint of population genetics and ecology.

SYNONYMY

The name *nigerrima* Leng has been applied to the melanic specimens found in the middle of the continent; *nigerrimoides* Hatch has been used for melanics from the Pacific Northwest. Both are synonyms of their respective subspecies.

Four specimens from Alberta (Taku, Medicine Hat, Burdett, and Cypress Hills) were identified as *denverensis* Casey. None of these are like my *denverensis* from the type locality, and all can be matched by specimens of *purpurea* from Manitoba. Hence *denverensis auctorum* (not Casey), is a synonym of *purpurea* (*s. lat.*).

RELATIONSHIPS

See under discussion of *limbalis* and *cimarrona* for details.

LOCALITIES

Map 9 indicates the positions of the localities listed below. Localities are listed for *purpurea* (*s. lat.*): NOVA SCOTIA, Sydney; QUEBEC, Hemingford; ONTARIO, Arnprior, Aylmer, Cedarville, Constance Bay, Credit Forks, Guelph, Hastings County, Hyde Park Corner, Kettleby, London, Mount Dennis, Ottawa, Rouge River, Sudbury, Trenton, Wasaga Beach, Willowdale; MANITOBA, Aweme, Brandon, Lyleton, Wawanesa; SASKATCHEWAN, Cut Knife, Gull Lake, North Battleford, Roche Percee, Saskatoon, Swift Current; ALBERTA, Bassano, Brooks, Burdett, Calgary, Castor, Cessford, Cypress Hills, Edmonton, Ghost River, Hanna, Lethbridge, Magrath, Medicine Hat, Merid, Orion, Pincher Creek, St. Mary's, Taber, Tilley, Walling; BRITISH COLUMBIA, south and southwest

Cicindela longilabris Say, 1824
(Nos. 66–69, Plate 3, and Map 10)

This species shares with *montana* Le Conte a bald, broadly excavated head, non-serrate elytral apices, and a long labrum. This combination of characters sets these species apart from all others which occur in North America. The dorsal surface of the elytra of *longilabris* is opaque, and on it there are many minute granules. In *montana*, the elytra are shining, and the dorsal surface is smooth, without granules.

GEOGRAPHICAL VARIATION AND SUBSPECIES

White markings of the elytra, though usually present, are lacking in a few individuals. The dorsal surface is black in specimens from the greater part of the mainland, and these represent the nominate subspecies, *longilabris longilabris* (Nos. 66–67, Plate 3). In eastern Labrador and on Newfoundland, this species is represented by *longilabris novaterrae* Leng (Nos. 68–69, Plate 3). Colour varies in the latter populations from black through bronze to green, perhaps depending on ecological conditions in different parts of the area.

SYNONYMY

Say's original description of *longilabris* is difficult to interpret. Although it is clear that he was referring either to *longilabris* as used here, or to *montana* Le Conte, the description neglects to mention the most important characters for distinguishing between these two species: lustre and presence or absence of minute granules on the elytra. As Say's type of this species has been destroyed, its identity must be established by means other than direct comparison.

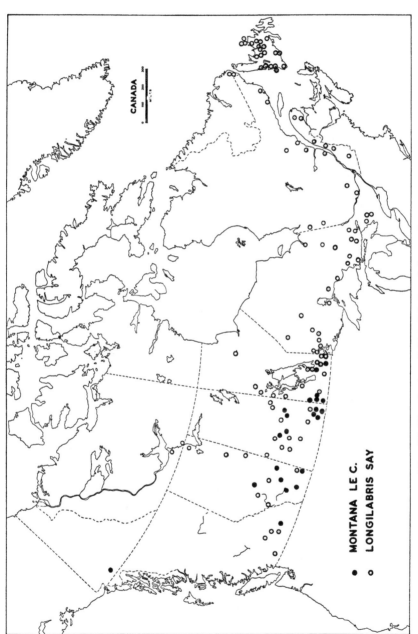

MONTANA LE C. ●

LONGILABRIS SAY ○

MAP 10. Distribution of *Cicindela longilabris* and *montana*

The type locality is given as "North West Territory," but this was a vast area including the Dakotas and extending to the Rocky Mountains, and covering areas occupied by both species under consideration. Thus it is important to determine from what part the original specimen of *longilabris* came. The specimen was evidently collected by an expedition under the direction of Major Long. Two were led by him, the first reaching as far west and north as Denver, Colorado. This would be within the range of *montana*. However, W. J. Brown of Ottawa has informed me that "When Say described the species from the other western trip (the first), he *always* gave the locality as Missouri or Missouri Territory, as Arkansas or as 'near the Rocky Mountains', when he did not make it more specific. He described nothing from 'North West Territories' in that paper."

From the above it seems certain that the types of *longilabris* were taken within the limits of the second Long expedition to the source of the St. Peter's River in 1823. In brief, the route went from Philadelphia, westward to Chicago, thence to Minneapolis, then to Big Stone Lake —the source of the St. Peter's River.

Big Stone Lake is on the boundary between the northeast corner of South Dakota and Minnesota and marks about the western limit of the expedition, as from this point the route led down the Red River to Lake Winnipeg and probably to what is now Victoria Beach or perhaps Fort Alexander. At any rate the expedition seems now to have followed the Winnipeg River to Lake of the Woods and thence along the north shore of Lake Superior across Lake Ontario and back home.

In much of this area the shiny black *montana* is not found, but the dull *longilabris* can probably be taken anywhere and is very abundant in much of it. Indeed, during September when the expedition was on its way home, the dull *longilabris* would be plentiful on the whole route from Lake Winnipeg to Lake Ontario. On the basis of probability the type locality for *longilabris* Say, therefore, is somewhere between Victoria Beach, Manitoba, and Lake Superior, or perhaps even farther east to Lake Ontario. The names, therefore, should be applied as at present.

In the original description of *albilabris*, Kirby (1837) mentions that the "elytra, under a powerful magnifier [are] covered with innumerable minute granules," and this statement is sufficient to establish that *albilabris* is the same as *longilabris*.

In 1924, Casey described *terracensis* as a subspecies of *oslari* based on a specimen collected at Terrace, British Columbia. I believe that *terracensis* is no more than a well-marked specimen of *longilabris*, based on Casey's original description: sub-opaque, deep black, quite well marked. I have several specimens of *longilabris* from widely separated localities which agree perfectly with this description.

RELATIONSHIPS

As was indicated above, this species and *montana* are very similar but do not seem to have any other close relatives in North America. Both have male genitalia of the same type, and a Palaearctic species, *sylvatica* Linneaus has markedly similar genitalia. These three species seem to form a natural unit.

HABITAT

Cicindela longilabris is a species found in sandy areas in or adjacent to coniferous forest. It abounds from Quebec to Alberta, especially where jack pine is found. It does not mix with *montana*, though sometimes they may be found occupying adjacent territory; for example, where a short grass meadow with heavier soil meets with the sandy soil on which a stand of conifers grows.

LOCALITIES

These are indicated on Map 10. I give here only the general distribution of the species, as follows: Newfoundland, Quebec, Ontario, Manitoba, Saskatchewan, Alberta, Northwest Territories, Yukon Territory, and British Columbia.

Cicindela montana Le Conte, 1868
(Nos. 60–65, Plate 3, and Map 10)

This species, like its close relative *longilabris*, is clearly isolated, morphologically, from any other North American tiger beetle. See under the discussion of *longilabris* for details. Its non-granulate elytra set *montana* apart from *longilabris*.

GEOGRAPHICAL VARIATION AND SUBSPECIES

The pattern of variation is complex, and the subspecies based on it must be regarded as tentative. In fact, Lindroth (1955, 19) denies the validity of several recognized here. Abundant material is needed for study, especially from the mountains of British Columbia.

Throughout the prairies is found the nominate subspecies, *montana montana* (Nos. 60–61, Plate 3). These populations comprise black, smooth individuals. Specimens collected at Aweme, Manitoba, are smaller than *m. montana*, and some individuals exhibit a metallic green tint on their elytra. The underside is usually much more brilliantly coloured than in the more western lowland individuals. This is the sub-species *montana spissitaris* Casey (No. 62, Plate 3), described from Aweme, Manitoba.

A third subspecies is found in the mountains of western North America, including British Columbia. The colour of the integument of this form varies from bronze through bronze-green to brilliant green, and the markings may be well developed, reduced, or absent. The surface of the dorsal integument may be relatively smooth or else coarsely sculptured. The montane populations may be arrayed in four subspecies on the basis of combinations of these characters.

Montana oslari Leng occurs throughout the more southern mountains of British Columbia. Its dorsal surface is either bronzy or green (No. 64, Plate 3), immaculate or with a reduced pattern of markings. In the same general area, though perhaps at different elevations, *montana laurenti* Schaupp is found. It is dark brown, feebly bronzed, and usually well marked with complete humeral and apical lunules and middle band. Sometimes a moderately long marginal line reaches the humeral lunule, but the most striking modification of the markings is the expansion of the apical lunule into a conspicuous elongate spot (No. 65, Plate 3). This pattern is subject to reduction. A third sub-species, *montana chamberlaini* Knaus (No. 63, Plate 3), occurs in British Columbia. It is bronzed, and the surface of the integument is coarsely sculptured. A fourth subspecies, of which I have seen no Canadian specimens, is vivid green with broad and distinct markings, and is called *montana perviridis* Schaupp. It is found in western British Columbia.

This classification is tentative as my understanding of the distribution pattern of the diagnostic characters is based on few specimens and few field observations. The relationships of these populations should be carefully investigated in the field, and special attention should be paid to intra-population variation and to the determination of ecological preferences.

SYNONYMY

In 1913, Casey described *canadensis*, from Calgary, Alberta, as a subspecies of *montana*. In 1914, he elevated this form to the status of species and described also *calgaryana*, from Lethbridge. Both of these forms are based on minor characters of sculpture, which are encompassed by the normal range of variation of *montana montana*. Another subspecies described by Casey, *ostenta columbiana*, from British Columbia, is founded on an inconstant colour character, and is a synonym of *montana perviridis* Schaupp.

HABITAT

Montana is a species of the short grass plains, but extending westward into the valleys and up into the mountains, and eastward through Saskatchewan and Manitoba to the Red Deer River, which seems to form its eastern limit. It prefers heavy clay soil, and is not found on sand.

RELATIONSHIPS

This topic is discussed under *longilabris*.

LOCALITIES

These are indicated on Map 10. I give here only the general distribution of the species: Manitoba, Saskatchewan, Alberta, British Columbia, and Yukon Territory.

Cicindela fulgida Say, 1823
(Nos. 77–82, Plate 4, and Map 11)

This species most closely resembles *parowana* Wickham, but the two may be easily separated by the characters given in the key. In both species the markings of the elytra and the colour of the pigmented areas are very similar to those of *lengi*. The markings are, however, heavier than in *lengi*, and the lustre of the elytra in *fulgida* and *parowana* has a more metallic quality (cf. Nos. 77–83, Plate 4, and Nos. 17–21, Plate 2).

GEOGRAPHICAL VARIATION AND SUBSPECIES

Size and colour exhibit slight variation, on the basis of which two subspecies may be recognized. The nominate subspecies, *fulgida fulgida*, is found in western United States and Alberta. The second subspecies, *fulgida westbournei* Calder, includes populations found in southern Saskatchewan and Manitoba.

Specimens of *f. fulgida* are usually over 12 mm. in length. They are brilliant coppery or occasionally greenish-coppery above, frequently with shining granules on the elytra. Usually the humeral lunule is rather widely separated at its tip from the middle band (Nos. 77–78, Plate 4).

Specimens assigned to *f. westbournei* rarely attain 12 mm. in length, 10.5 mm. being about average. Typically the colour is dark brown tinged with red, green, or purple, but with little brilliance and a greasy appearance. The humeral lunule approaches the middle band closely, often joining it (Nos. 79–82, Plate 4). Nine specimens collected south-

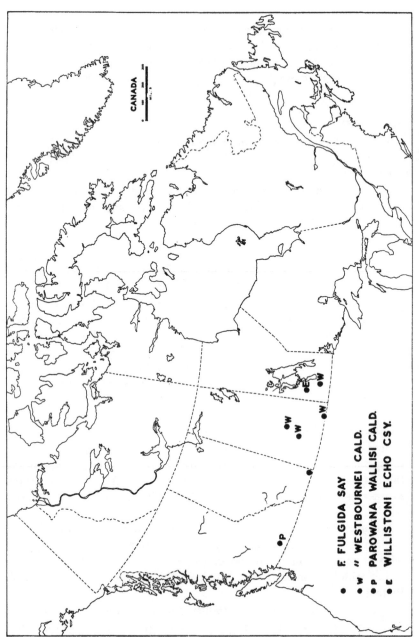

MAP 11. Distribution of *Cicindela fulgida*, *parowana wallisi*, and *willistoni echo*

east of Blucher, Saskatchewan, are definitely more shining and more brilliantly coloured green or blue than are the Westbourne specimens. None is of the bright reddish-cupreous of *f. fulgida*, nor is any of these specimens as large. Possibly the two subspecies recognized here are connected by a series of populations with intermediate characteristics.

SYNONYMY

Cazier (1936) placed *wallisi* Calder as a synonym of this species. However, according to Cazier (*in litt.*) *fulgida* does not enter the Great Basin, and Cazier now refers *wallisi* to *parowana* Wickham.

The subspecies *f. westbournei* was originally named *f. elegans* Calder (1922). Discovering that this name was preoccupied, the author gave this subspecies the name used here.

RELATIONSHIPS

This species and *parowana* Wickham are closely allied, as is indicated by marked similarity in the structure of the male genitalia, as well as in external characters. These two species do not appear to be closely related to any other tiger beetle species.

HABITAT

Both *Cicindela fulgida* and *parowana* live on saline or alkaline soil. The type locality of *f. westbournei* is a small plain at Westbourne, Manitoba, with many bare alkaline patches. At Onefour, Alberta, *f. fulgida* was taken among sparse vegetation along the banks of the Lost River, on dry, alkaline clay.

LOCALITIES

The positions of the localities listed below are indicated on Map 11.

fulgida fulgida Say: ALBERTA, Onefour, 49° 01′ 00″ N., 110° 20′ 00″ W.

fulgida westbournei Calder: MANITOBA, Westbourne; SASKATCHEWAN, southeast of Blucher, Elbow, Roche Percee

Cicindela parowana Wickham, 1905
(No. 83, Plate 4, and Map 11)

This species is most similar to *fulgida* Say, but can be confused with *lengi* W. Horn. See the discussion under *fulgida* for a summary of diagnostic characters.

SUBSPECIES

In Canada this species is represented by the subspecies *parowana wallisi* Calder. This population was originally called *azurea* Calder, but the author discovered that this name was preoccupied, and changed it to *wallisi*.

RELATIONSHIPS

Details pertinent to this species are presented in the discussion under *fulgida*.

HABITAT

This species prefers alkaline soil. As an indication of how closely some of the tiger beetles are confined to particular types of soil, *parowana wallisi* was taken in 1919 on a small path extending across alkaline soil, toward lower Okanagan Lake, south of Penticton, British Columbia. Thirty years later, it was possible to give directions to a party of collectors who found the colony within a few yards of its previous location!

LOCALITY

In Canada, this species is known only from the type locality of *p. wallisi*: between Penticton and lower Okanagan Lake.

Cicindela lengi W. Horn, 1908
(Nos. 17–21, Plate 2, and Map 12)

This species is similar in appearance to *formosa formosa* Say, but the characters given in the key are sufficient for purposes of distinguishing between the two. See also the discussion of the diagnostic features of *fulgida*, under that species.

GEOGRAPHICAL VARIATION AND SUBSPECIES

The sole representative of this species in Canada is the subspecies *lengi versuta* Casey (Nos. 19–21, Plate 2), of which I have examined 350 specimens.

Bright blue ventral and lateral thoracic sclerites characterize the populations of the nominate subspecies, *lengi lengi* (Nos. 17–18, Plate 2). However, in a series of forty-six specimens from Colorado and Kansas representing populations referred to *l. lengi*, three have dull coppery thoracic sclerites, three have brilliant coppery ones, and three have yellowish-green.

Specimens from the type locality of *lengi versuta* (Aweme, Manitoba) have brilliant coppery thoracic sclerites. This phenotype is the dominant one throughout western Canada. In Alberta, a few specimens have

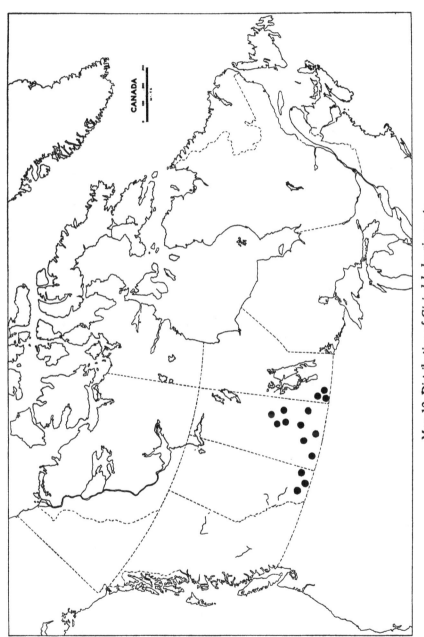

CANADA

Map 12. Distribution of *Cicindela lengi versuta*

yellowish-green, yellowish, or steel-grey thoracic sclerites, but none have them blue or blue-green.

The colour of the dorsal surface exhibits some variation. Western specimens have a slight greenish sheen. Indeed, some from Alberta are green rather than coppery red above, and two from Medicine Hat, Alberta, and one from Chaplin, Saskatchewan, are brilliant blue. One male, from White Fox, Saskatchewan, taken mating with the usual red form, is black.

SYNONYMY

The subspecies *lengi lengi* has been recorded from Alberta and Saskatchewan by Mrs. Vaurie (1950). However, the phenotype which is typical of the nominate subspecies is relatively infrequent in the populations which occupy this area, so *lengi lengi* should not be regarded as being here. It is possible that the Alberta populations represent a "blend" between the two subspecies.

RELATIONSHIPS

Although this species and *ancocisconensis* Harris are readily distinguishable on the basis of external characteristics, the males of both species have genitalia which are markedly similar. I believe these two are closely related. However, they are not very close to any other extant species of *Cicindela*.

HABITAT

Cicindela lengi seems to require dry sand. This species is found on sand dunes in the southern portions of the prairies as well as along jack pine ridges in the boreal forest. *Limbata* and *formosa* are its closest associates from an ecological standpoint.

LOCALITIES

Those listed here are indicated on Map 12: MANITOBA, Aweme (type locality, Spruce Woods Forest Reserve), Oak Lake, St. Lazare; SASKATCHEWAN, Beaver Creek, 9 miles northeast of Canora, Cut Knife, Dundurn, Gascoigne, Good Spirit Lake, Holbein, Indian Head, Lake Chaplin, Lake Manitou, Nipawin, Pike Lake, Prince Albert, Pas Trail, Sceptre, Torch River (White Fox); ALBERTA, Claysmore, Edgerton, Edmonton, Empress, Lethbridge, Lost River, Medicine Hat, Fort MacLeod, Orion

Cicindela ancocisconensis Harris, 1852
(No. 39, Plate 2, and Map 1)

The colour of the dorsal integument and the form of the elytral markings are sufficient to distinguish this species from other tiger beetles which occur in the northeast.

RELATIONSHIPS

The closest relative of *ancocisconensis* is *lengi*. For details, see discussion under the latter species.

LOCALITIES

In Canada, this species is known only from Gaspé County, Quebec. This locality is indicated on Map 1.

Cicindela tranquebarica Herbst, 1806
(Nos. 51–59, Plate 3, and Map 13)

The long, obliquely directed descending arm of the humeral lunule is sufficient to distinguish this species from all others found in Canada. Next to *repanda*, the commonest tiger beetle in our area is probably *tranquebarica*.

GEOGRAPHICAL VARIATION AND SUBSPECIES

Colour and markings vary geographically and ecologically. The geographical pattern is complex and is poorly understood. Therefore, even though a number of names have been proposed for subspecies, they should not be used. In many eastern specimens, the dorsal surface is blackish, and the markings are usually slender (No. 51, Plate 3). This form is the nominate subspecies, *t. tranquebarica*. This phenotype occurs also in western Canada, but more rarely. The predominant phenotype in the prairie provinces, but also occurring in eastern Canada, has been called *tranquebarica kirbyi*. In this form, the markings are very broad (No. 56, Plate 3) and the elytra are brownish or bronzy, westward becoming greenish. This form seems to thrive on alkaline soil and on clay.

Rarely in the west, a form has been taken with the markings of *t. tranquebarica*, but of considerably smaller size. This is called *tranquebarica minor* Leng (No. 52, Plate 3).

Farther north is found *tranquebarica borealis*. This phenotype is differentiated from *t. tranquebarica* and *t. kirbyi* by its humeral lunule, which is broken or at least very narrow in the middle. At Fort Smith,

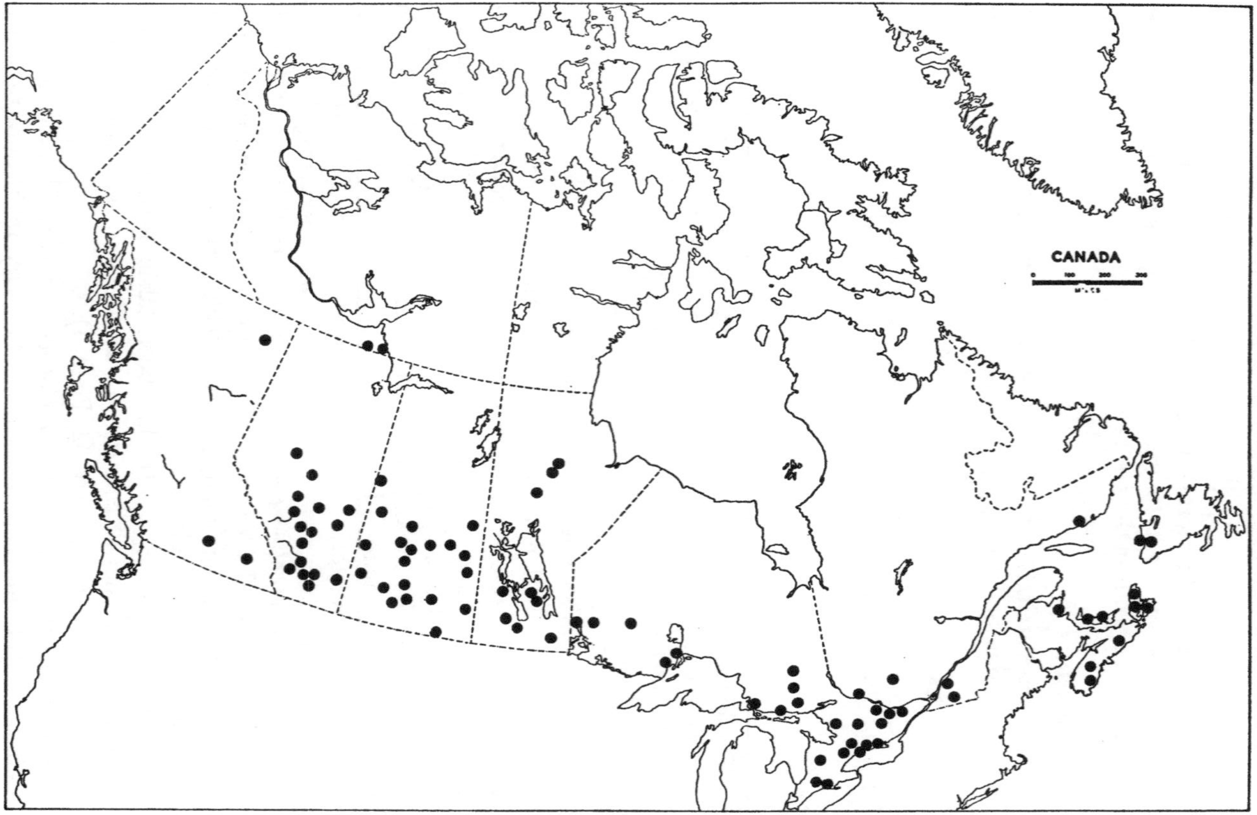

MAP 13. Distribution of *Cicindela tranquebarica*

Northwest Territories, both *t. borealis* and *t. kirbyi* were found. The former occupied sandy areas, whereas *kirbyi* was restricted to an area where the soil was heavy and alkaline. This suggests either that development of the markings is influenced by soil composition, or that different ecotypes which can be recognized by their elytral markings occur within this species. A third alternative is that these types are sibling species. These possibilities should be checked by field observations and experiments.

In British Columbia, specimens occur which have green integuments and much reduced elytral markings. Such individuals represent *t. vibex* Horn (No. 59, Plate 3).

SYNONYMY

All of the trinomials listed above are considered as synonyms here. They should not be applied until the underlying causes of the observed geographical and ecological variation are understood.

Hatch (1953) differentiates *t. roguensis* from *t. vibex* only on an unstable colour characteristic—dark green *versus* bright green. This name is also a synonym of *tranquebarica*.

A subspecies described by Leng, *t. horiconensis*, occurs in the northeast. However, it does not seem to be different from *t. kirbyi*.

RELATIONSHIPS

The closest relative of this species appears to be the Californian species *latesignata* Le Conte, for the two have markedly similar genitalia.

HABITAT

As indicated above, *tranquebarica* is at home both on clay, alkaline or otherwise, and on sandy soil.

LOCALITIES

These are listed here, and are indicated on Map 13: NOVA SCOTIA, distributed to Cape Breton; NEWFOUNDLAND, south of Stephenville Crossing (St. George Bay region); PRINCE EDWARD ISLAND, Brackley Beach, Dalvay House; NEW BRUNSWICK, Tabusintac; QUEBEC, Chilcott Lake, Fort Coulonge, Godbout, Ile Perrot, Kazabazua, Knowlton, Mount Lyall, Natashquan (Labrador), Partageville, Ste Anne, Stoke; ONTARIO, Arnprior, Algonquin Park, Algoma, Attawapiskat, Bird's Creek (Hastings County), Bond Lake, Britannia, Cedarville, Constance Bay, Credit Forks, Hyde Park Corner, Kettleby, Lake Nipissing, Lake Wilson, Leskard, London, Low Beach, Macdiarmid, Malachi, Muskoka, Musselman's Lake, Nipigon, Ottawa, Port Hope, Port Sidney, Sault Ste Marie, Sioux Lookout, Sudbury, Smoky Falls on Mattagami River, Temagami, Toronto, Wasaga; MANITOBA, Aweme, Gillam, Kelwood, Marchand, Mile 214 and Mile 256 Hudson Bay Railway, Shoal Lake, Victoria Beach, Westbourne; SASKATCHEWAN, Battle River, Beaver Creek, 9 miles northeast of Canora, Cut Knife, Dafoe, Dundurn, Elbow, Estevan, Gascoigne, Good Spirit Lake, Great Sand Hills, Gull Lake, Indian Head, Kindersley, Lake Chaplin, Maryfield, Pas Trail, Saskatoon, Swift Current, Three-mile Creek, Torch River, Tunstall; ALBERTA, Barnwell, Bilby, Calgary, Castor, Chin, Claresholm, Clymont, Consort, Edmonton, Fawcett, High River, Lethbridge, Lundbreck, Fort MacLeod, Medicine Hat, Red Deer, Rosedale, St. Mary's, Simpson, Soda Lake, Stavely, Tofield, Wetaskiwin; BRITISH COLUMBIA, southwest; NORTHWEST TERRITORIES, salt plain west of Fort Smith, Fort Norman

Cicindela willistoni Le Conte, 1879
(No. 76, Plate 3, and Map 11)

The bald head, serrulated elytral apices, rather rough elytral sculpture, brown-bronze colour above, markings well developed and consisting of complete humeral and apical lunules and a complete middle band, combine to form a group of characters by which this species can be recognized easily.

SUBSPECIES

The subspecies reported from Canada is *willistoni echo* Casey. Two specimens were collected, supposedly at Kelwood, Manitoba. Now, this species is found on the saline shores of Great Salt Lake, Utah. That *willistoni* should occur so far away, and in an entirely different environment, seems scarcely credible. Therefore, I doubt the validity of this record for Canada.

RELATIONSHIPS

The complex internal sac of the male genitalia is intermediate between that of *tranquebarica* and that which is seen in the group of species to which *punctulata* belongs. However, *willistoni* does not appear to be directly related to either group.

LOCALITY

Kelwood, Manitoba, which is indicated on Map 11.

Cicindela punctulata Olivier, 1790
(No. 96, Plate 4, and Map 14)

Little difficulty should be experienced in the identification of this species, the characters in the key being sufficient. In addition, it may be noted that the markings comprise a few white dots.

VARIATION

The only subspecies of this species to occur in Canada is *p. punctulata*. Completely or almost completely immaculate individuals are not rare, and specimens occur in which an interrupted middle band, marginal band, and humeral lunule are present and in which the apical lunule is complete.

SYNONYMY

The name *boulderensis* Casey, 1909, was given to the form of *punctulata* which occurs in Colorado and Manitoba, but the supposed differences between this phenotype and typical *punctulata* do not justify the retention of this name.

RELATIONSHIPS

The structure of the male genitalia indicates that this species is related to a complex of southwestern species, including *flavopunctata* Chevrolat, *obsoleta* Say, and several others.

HABITAT

This species seems to prefer dry sandy loam, preferably hard packed. It is found abundantly on paths and roads. Larvae may be found in burrows among the grass tufts along the edges. *Punctulata* is also found on dry, sparsely grassed prairies.

LOCALITIES

These are listed here, and are indicated on Map 14: ONTARIO, Bell's Corners (near Ottawa), eastern Ontario, Guelph, Hyde Park Corner, Kettleby, Leskard, London; MANITOBA, Aweme, Hartney, Treesbank; SASKATCHEWAN, Estevan; ALBERTA, Burdett, Happy Valley, Lethbridge, Medicine Hat

Cicindela pusilla Say, 1817
(Nos. 97–102, Plate 4, and Map 15)

Some specimens of this species may be confused with *punctulata*. However, the row of large punctures so characteristic of the latter is lacking in *pusilla*.

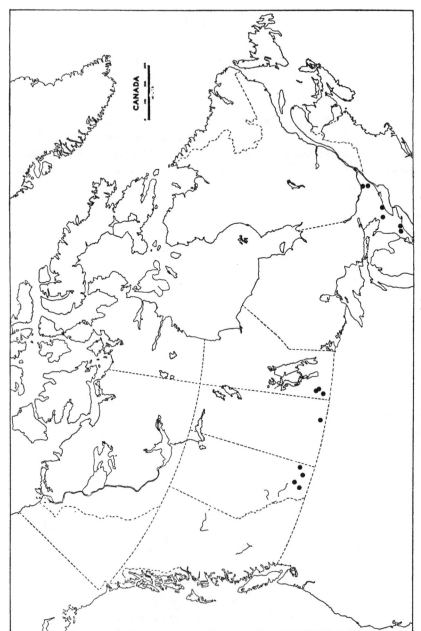

MAP 14. Distribution of *Cicindela punctulata punctulata*

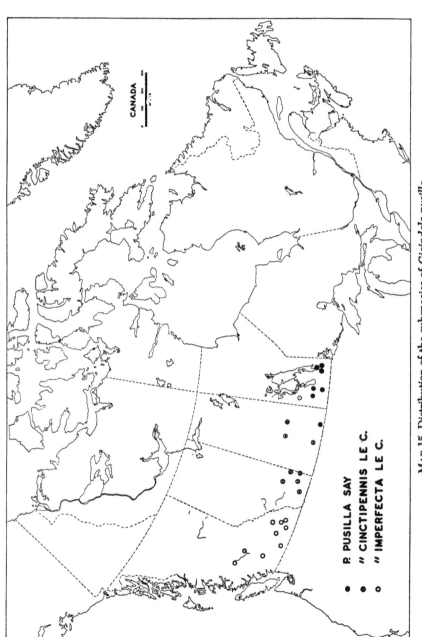

P. PUSILLA SAY ●

" CINCTIPENNIS LE C. ◉

" IMPERFECTA LE C. ○

MAP 15. Distribution of the subspecies of *Cicindela pusilla*

GEOGRAPHICAL VARIATION AND SUBSPECIES

Colour, pattern of markings, and sculpture of the elytra vary; three subspecies are based on these characteristics.

In Manitoba the populations consist entirely of specimens with the dorsal surface coloured black, and with highly variable maculation. The elytra are covered with "scattered, very minute punctures, which are oblique, as if formed by a pointed instrument directed toward the anterior part of the insect so that the surface before each puncture is a little elevated" (Say (1824), description of *terricola*, a synonym of *pusilla*). Among the punctures, the elytral surface is so minutely reticulate as to be almost sericeous. Specimens with these characteristics are placed in the subspecies *pusilla pusilla* (Nos. 97–100, Plate 4).

In Alberta, specimens of *pusilla* are invariably green with well-developed elytral markings; small fovea-like punctures, green or blue at the bottom, are very evident. The microreticulation between the punctures is well marked and the surface is metallic. These specimens represent the subspecies *pusilla cinctipennis* Le Conte (No. 102, Plate 4).

I have seen a population sample collected at Roche Percee, Saskatchewan, comprising several hundred individuals. This sample exhibits a wide range of variation. It includes black individuals with all types of marking patterns, and with the same kind of sculpture characteristic of *p. pusilla*. Heavily marked and immaculate bronze, blue, and green individuals are also present, and these have the fovea-like punctures characteristic of *p. cinctipennis*. Thus this population includes the variants of both subspecies, and in this sense it is intermediate.

The third subspecies, *pusilla imperfecta* Le Conte, from British Columbia, is a dark greenish form with elytral colouring and sculpture much like those of *p. cinctipennis*. However, the pattern of the elytra differs considerably from that seen in *p. cinctipennis*: the greater part of the marginal band has disappeared leaving at most a spur from the humeral lunule joining it to the long middle band. The transverse arm of the middle band commences far from the margin and is very short. In a few specimens the pattern is much reduced but it always shows indications that it is derived from the pattern described here (No. 101, Plate 4).

RELATIONSHIPS

The very complex sclerites of the internal sac of the male genitalia of this species are also present in *cuprascens*, *nevadica*, and *lepida*, as well as in a number of other species not treated here. These species constitute a natural division of *Cicindela*, but just how they are related to one another is not established.

HABITAT

Alkaline or saline situations with sparse vegetation and bare spots are the favourite haunts of at least the prairie populations of this species. In eastern British Columbia, *pusilla imperfecta* has been taken on sandy clay soil along river banks.

LOCALITIES

Those which are listed below are indicated on Map 15.

pusilla pusilla Say: MANITOBA, Aweme, Baldur, Beulah, Bird's Hill, Stony Mountain, Westbourne, Winnipeg; SASKATCHEWAN, southeast of Blucher, Roche Percee

pusilla cinctipennis Le Conte: SASKATCHEWAN, Roche Percee, Rock-glen, Rudy, Saskatoon, St. Victor, Willow Bunch; ALBERTA, Drumheller, Empress, Lethbridge, Medicine Hat, Redcliff

pusilla imperfecta Le Conte: ONTARIO, eastern Ontario (almost certainly in error); MANITOBA, Kelwood (almost certainly in error); BRITISH COLUMBIA, Athalmer, Cawston, Canal Flats, Chilcotin, Clinton, Cranbrook, Creston, Fort Steele, Lillooet, Okanagan Falls, Oliver, Wasa Lake, White Lake

Cicindela cuprascens Le Conte, 1852
(No. 105, Plate 4, and Map 16)

The identification of this species should present no difficulty. The peculiarly shaped humeral lunule separates it from all of our forms save *hirticollis* and *nevadica knausi*. However, the large white spot between the suture and the shoulder, present in *cuprascens*, is absent in the other two species (cf. No. 105, Plate 4, Nos. 49–50, Plate 3, and Nos. 103–104, Plate 4). In addition, it may be noted that, in *cuprascens*, the markings are quite strongly embossed.

RELATIONSHIPS

This topic is discussed under *pusilla*.

HABITAT

Some 16 miles east of Aweme, Manitoba, is a deep deposit of pure sand, quite close to the Assiniboine River. Springs welling out from the sand have caused a recession of the high sand wall for a con-

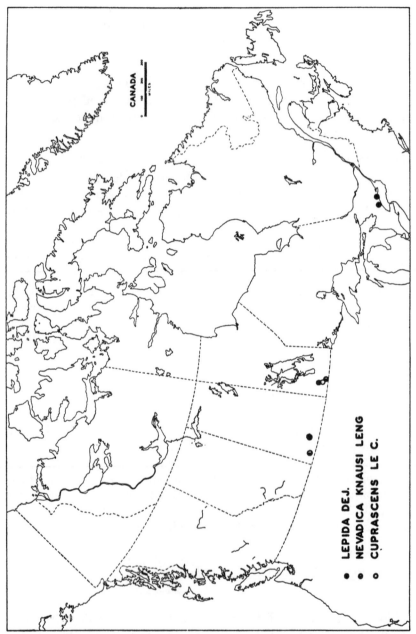

MAP 16. Distribution of *Cicindela cuprascens*, *nevadica*, and *lepida*

LEPIDA DEJ.

NEVADICA KNAUSI LENG

CUPRASCENS LE C.

CANADA

MILES
0   100   200   300

siderable distance from the river, thus forming a ravine. There is now a wide, semi-circular area, locally known as the "Punch Bowl," formed at the inner end of the long ravine. On top and along the slope of the Punch Bowl an occasional *cuprascens* is taken, at intervals of several years. Search for this species along the river and in other apparently suitable situations has been unavailing. Perhaps, then, this species is not a permanent resident at this site, but only an occasional visitor.

LOCALITY

About 16 miles east of Aweme. This spot is indicated on Map 16.

Cicindela nevadica Le Conte, 1875
(Nos. 103–104, Plate 4, and Map 16)

This species is represented in Canada by the subspecies *nevadica knausi* Leng. In general appearance, this subspecies resembles *cuprascens*, but it is smaller, darker above, and lacks the basal white dot on the elytra.

RELATIONSHIPS

This topic is discussed under *pusilla*.

HABITAT

This species lives on wet mud, along the margins of saline or alkaline lakes and streams.

LOCALITIES

These are indicated in Map 16, and are listed here: MANITOBA, on shore of Watson's Lake, about six miles south of Hilton; SASKATCHEWAN, Coteau Lake

Cicindela lepida Dejean, 1831
(No. 106, Plate 4, and Map 16)

In addition to the characters presented in the key, it may be noted that the labrum of *lepida* is one-toothed, and the legs, antennae, and elytra are very pale.

SYNONYMY

Casey described as *insomnis* a form from Kansas with head and prothorax a pure, brilliant green. This form occurs occasionally in the Aweme sand hills. It appears to be no more than a colour form, perhaps due to immaturity.

RELATIONSHIPS

For details, see discussion under *pusilla*.

HABITAT

This species lives on pure white or pale yellow sand, and is found away from vegetation. It is one of the most perfectly camouflaged of tiger beetles, and is almost impossible to detect until it moves.

LOCALITIES

These are indicated on Map 16, and are listed here: ONTARIO, Leskard, Port Hope; MANITOBA, sand hills, near Aweme; SASKATCHEWAN, southern Saskatchewan (Vaurie, 1950)

V. Check-List of the Species and Subspecies of the Cicindelidae of Canada

Omus Eschscholtz, 1829
  Omus (Megomus) dejeani Reiche, 1838
  Omus (Omus) californicus Eschscholtz, 1829
    Omus (Omus) californicus audouini Reiche, 1838
Cicindela Linnaeus, 1758
  Cicindela repanda Dejean, 1825
    Cicindela repanda repanda Dejean, 1828
    Cicindela repanda novascotiae Vaurie, 1951
  Cicindela duodecimguttata Dejean, 1825
  Cicindela oregona Le Conte, 1857
    Cicindela oregona guttifera Le Conte, 1857
    Cicindela oregona oregona Le Conte, 1857
    Cicindela oregona depressula Casey, 1897
  Cicindela hirticollis Say, 1817
  Cicindela limbata Say, 1823
    Cicindela limbata nympha Casey, 1913
    Cicindela limbata hyperborea Le Conte, 1863
  Cicindela sexguttata Fabricius, 1775
    Cicindela sexguttata sexguttata Fabricius, 1775
    Cicindela sexguttata denikei Brown, 1934
  Cicindela patruela Dejean, 1825
  Cicindela scutellaris Say, 1823
    Cicindela scutellaris scutellaris Say, 1823
    Cicindela scutellaris lecontei Haldeman, 1853
  Cicindela formosa Say, 1823
    Cicindela formosa generosa Dejean, 1831
    Cicindela formosa manitoba Leng, 1902
    Cicindela formosa gibsoni Brown, 1940
    Cicindela formosa formosa Say, 1817
  Cicindela cimarrona Le Conte, 1868
  Cicindela decemnotata Say, 1817
  Cicindela limbalis Klug, 1834

Cicindela purpurea Olivier, 1790
  Cicindela purpurea purpurea Olivier, 1790
  Cicindela p. purpurea x *p. auduboni* Le Conte, 1845
  Cicindela purpurea pugetana Casey, 1914
Cicindela longilabris Say, 1824
  Cicindela longilabris novaterrae Leng, 1902
  Cicindela longilabris longilabris Say, 1824
Cicindela montana Le Conte, 1861
  Cicindela montana montana Le Conte, 1861
  Cicindela montana spissitarsis Casey, 1913
  Cicindela montana laurenti Schaupp, 1883
  Cicindela montana oslari Leng, 1902
  Cicindela montana chamberlaini Knaus, 1925
  Cicindela montana perviridis Schaupp, 1884
Cicindela fulgida Say, 1823
  Cicindela fulgida fulgida Say, 1823
  Cicindela fulgida westbournei Calder, 1922
Cicindela parowana Wickham, 1905
  Cicindela parowana wallisi Calder, 1922
Cicindela lengi W. Horn, 1908
  Cicindela lengi versuta Casey, 1913
Cicindela ancocisconensis Harris, 1852
Cicindela tranquebarica Herbst, 1806
Cicindela willistoni Le Conte, 1879
  Cicindela willistoni echo Casey, 1879
Cicindela punctulata Olivier, 1790
  Cicindela punctulata punctulata Olivier, 1790
Cicindela pusilla Say
  Cicindela pusilla pusilla Say, 1817
  Cicindela pusilla cinctipennis Le Conte, 1848
  Cicindela pusilla imperfecta Le Conte, 1851
Cicindela cuprascens Le Conte, 1852
Cicindela nevadica Le Conte, 1875
  Cicindela nevadica knausi Leng, 1902
Cicindela lepida Dejean, 1831

LITERATURE CITED

CASEY, T. L.
1909. Studies in the Caraboidea and Lamellicornia. Canadian Entomologist, 41: 253–284.
1913. Studies in the Cicindelidae and Carabidae of America. Memoirs on the Coleoptera. New Era Publishing Co., Lancaster, Pa. Vol. IV, pp. 1–192.
1914. Miscellaneous notes and new species. *Ibid.*, vol. V, pp. 355–378.
1924. Additions to the known Coleoptera of North America. *Ibid.*, vol. XI, 347 pp.

CAZIER, M. A.
1936. Review of the *willistoni, fulgida, parowana* and *senilis* groups of the genus *Cicindela* (Coleoptera–Cicindelidae). Bulletin of the Southern California Academy of Sciences, 35(3): 156–163.

CHAPMAN, R. N.; MICKEL, C. E.; MILLER, G. E.; PARKER, J. R.; and KELLEY, E. G.
1926. Studies in the ecology of sand dune insects. Ecology, 7: 416–427, 549–557.

CRIDDLE, N.
1907. Habits of some Manitoba tiger beetles (*Cicindela*). Canadian Entomologist, 39: 105–114.
1910. Habits of some Manitoba tiger beetles (*Cicindela*). No. 2. *Ibid.*, 42: 9–15.
1919. Fragments of the life habits in Manitoba insects. *Ibid.*, 51: 97–101.
1925. A new *Cicindela* from Alberta. *Ibid.*, 57: 127–128.

FRICK, K. E.
1957. Biology and control of tiger beetles in alkali bee nesting sites. Journal of Economic Entomology, 50: 503–504.

HATCH, M. H.
1953. The beetles of the Pacific Northwest. Pt. I. Introduction and Adephaga. University of Washington Press, Seattle, Wash. vii + 340 pp., 37 plates.

HORN, W.
Coleoptera. Adephaga. Family Carabidae, Subfamily Cicindelinae. Genera Insectorum dirigés par P. Wytsman. Louis Desmet-Verteneuil, Bruxelles. Fascicules 82 A, 1908, 82 B, 1910, 82 C, 1915, pp. 1–486, plates 1–23.
1928. Notes on the tiger beetles of Minnesota. Technical Bulletin 56, Minnesota Agricultural Experiment Station, pp. 9–13.
1930. Notes on the races of *Omus californicus* and a list of the Cicindelidae of America north of Mexico. Transactions of the American Entomological Society, 56: 73–86, Plate VII, figs. 1–25.

KIRBY, W.
1837. Insects. Coleoptera. *In* Richardson, Fauna boreali-americana; or the zoology of the northern parts of British America. Vol. 4, 219 pp. Norwich.

LENG, C. W.
1920. Catalogue of the Coleoptera of American north of Mexico. Sherman, Mount Vernon, N.Y. 470 pp.
1927. Supplement to Catalogue of the Coleoptera of America north of Mexico. Sherman, Mount Vernon, N.Y. 78 pp.

LINDROTH, CARL H.
   1955. The carabid beetles of Newfoundland including the French islands St.
      Pierre and Miquelon. Opuscula Entomologica, Supplementum XII, pp.
      1–60, 58 figs.
MAYR, E.
   1942. Systematics and the origin of species. Columbia University Press, New
      York. 334 pp.
MAYR, E., LINSLEY, E. G., and USINGER, R. L.
   1953. Methods and principles of systematic zoology. McGraw-Hill Book Co.,
      Inc., New York. 328 pp., 45 figs.
PAPP, HELGA
   1952. Morphologische und phylogenetische Untersuchungen an *Cicindela*-
      Arten. Unter besonderer Berücksichtigung der Ableitung der neark-
      tischen Formen. Oesterreichische Zoologische Zeitschrift, 3(5): 494–533,
      figs. 1–12.
RIVALIER, E.
   1954. Démembrement du genre *Cicindela* Linne. II. Faune américaine. Revue
      française d'Entomologie, 21: 249–268, figs. 1–8.
SAY, T.
   1824. Coleoptera. *In* Keating, W. H. Narrative of an expedition to the source
      of St. Peter's River, Lake Winnepeek, Lake of the Woods, etc., etc.,
      performed in the year 1823 by order of the Hon. J. C. Calhoun, Secre-
      tary of War, under the command of Stephen H. Long, Major U.S.T.E.
      H. C. Carey and I. Lea, Philadelphia, vol. II, pp. 268–378.
SCHAUPP, F. G.
   1883. Synoptic tables of Coleoptera. Cicindelidae. Bulletin of the Brooklyn
      Entomological Society, 6: 73–108.
SHELFORD, V. E.
   1908. Life history and larval habits of the tiger beetles. The Journal of the
      Linnean Society of London. Zoology, pp. 157–184.
   1913. Life history of a bee fly. Annals of the Entomological Society of
      America, 6: 213–225.
   1917. Color and color pattern mechanism of tiger beetles. Illinois Biological
      Monographs, Nov. 1917, 134 pp.
SMYTH, E. G.
   1935. Analysis of *C. purpurea* group. Entomological News, 46: 14–19, 44–49.
VAURIE, P.
   1950. Notes on the habitats of some North American tiger beetles. Journal of
      the New York Entomological Society, 58: 143–153.
   1951. Five new subspecies of tiger beetles of the genus *Cicindela* and two
      corrections (Coleoptera, Cicindelidae). American Museum Novitates,
      1479: 1–12.

INDEX

ALL names marked with an asterisk are synonyms or homonyms. When more than one reference is given, the page on which the description or diagnosis appears is indicated by italic numerals.

Lightning Source UK Ltd.
Milton Keynes UK
UKHW012358200722
406167UK00001B/300